ANATOMY AN̲ ͟ ͟ ͟ ͟ ͟ ͟ ͟ ͟ ͟ ͟
for Nurses

ANATOMY AND PHYSIOLOGY for Nurses

10th Edition

Roger Watson BSc, PhD, RGN, CBiol, MIBiol
Senior Lecturer
Department of Nursing Studies,
The University of Edinburgh,
Edinburgh, UK

BAILLIÈRE TINDALL
London Philadelphia Toronto Sydney Tokyo

Baillière Tindall
An imprint of Harcourt Brace and Company Limited

© 1995 Baillière Tindall
© Harcourt Brace and Company Limited 1998

This book is printed on acid-free paper

First published 1939
Ninth edition 1979
Tenth edition 1995, all previous editions were published in the Nurses Aids Series
Reprinted 1998

ELBS edition 1979
(ISBN 07020 0738 2)
Sinhala edition (Official Language Affairs, Colombo) 1958
Turkish edition (Turkish Government) 1960
Spanish edition (C. E. C. S. A., Mexico) 1973
Hindi edition (N. R. Brothers, Indore) 1978

A catalogue record for this book is available from the British Library

ISBN 0-7020-1801-5

Printed in China
NPCC/02

CONTENTS

Preface

Producing the Tenth Edition of *Anatomy and Physiology for Nurses* has been a privilege. The previous editions have been marked by their consistency in presenting material at a level which ensures the utility of the book by a wide readership in nursing and other professions allied to medicine. It is sincerely hoped that this tradition has been maintained.

This edition is marked by a substantial reorganization of the contents into six sections. Each of the sections covers one of the major concepts in anatomy and physiology and is the method of organizing the systems of the body which is favoured by the majority of current textbooks in anatomy and physiology. Organizing the systems of the body in this particular way helps the student to understand the major function of a system and to make the appropriate links with other systems. For example, the skeletal system and the muscular system are both concerned with the posture and movement of the body. In order to achieve this there are anatomical links between the two systems and co-ordination of the two systems in order to achieve both posture and movement. It should be noted, however, that both the skeletal system and the muscular system have other functions, such as calcium homeostasis in the case of the skeleton and generation of heat by the muscular system, and this is true of most of the systems described in this text. These functions are described under other concepts, for example, calcium homeostasis is also a function of the endocrine system and this is described in the section Control and Coordination.

Additional material has been introduced sparingly. However, there are notable additions in the following chapters: *The Chemical Level of Organization; The Tissues, Organs, Systems and Homeostasis*; and *Non-specific Resistance and the Immune System*. Scattered throughout are some minor enhancements such as the inclusion of the action potential (Chapter 5), the neuromuscular junction (Chapter 13), and the electrocardiograph (Chapter 16).

Alterations to illustrations have led to a number of figures being enlarged slightly and discrete subsections of information have been highlighted in boxes, which are an entirely new featue of the text. In addition, each chapter is introduced in such a way that the reader is clear about what the subject material is and each section is introduced in order to explain the links between the systems that are covered.

The text concludes with a short bibliography of recent books which have been consulted in the preparation of this edition and which should prove useful to the reader for further reference.

Roger Watson
August 1995

SECTION I: THE CHARACTERISTICS OF LIVING MATTER

Section Introduction

The first section of this text includes material which is fundamental to an understanding of subsequent sections. The material is presented in a logical way, beginning with the simplest level of organization of matter up to the most complex level, found in the human body, of tissues, organs and systems.

The chemical level of organization considers the strong bonds which occur between atoms in the formation of molecules, these are the covalent bonds. Such bonds between carbon and hydrogen are the basis of the building of hydrocarbon molecules, which represent the basic structure of all biological macromolecules—the lipids, carbohydrates, proteins and nucleic acids. Weak interactions, principally the hydrogen bond, which is an essential feature of the structure and function of proteins and nucleic acids, are also considered.

At the cellular level of organization the plasma membrane, which contains the cellular contents and also controls the intracellular environment, is considered. The transport processes which work to exchange molecules and water between the extracellular and intracellular environments are described and the processes whereby cells divide to produce either identical daughter cells or gametes for subsequent involvement in sexual reproduction are introduced.

Finally, the highest levels of organization, which involve the organization of cells into tissues, tissues into organs and organs into systems, are presented. Before leaving this section the concept of homeostasis, which maintains the internal environment of the organisms, is introduced. Homeostasis is a unifying concept in physiology as it involves all of the systems working together to respond to changing circumstances and stresses imposed upon the organism by changes in the external environment or, in some cases, by disease.

Chapter 1
Introduction

Nurses care for human beings in illness and during vulnerable periods of their lives. In order to do this properly they need to understand the structure and function of the body and to be able to relate this knowledge to the care of the patient. Each organ in the body plays its part in maintaining the health of the whole, and if one organ is at fault the whole body will be affected. The structure of each part suggests the function, and the function suggests the structure, so that the study of the human body is a logical process of thinking and reasoning, not merely of memorizing.

Terms

1. *Anatomy* is the study of the structure of the body
2. *Physiology* is the study of its function

Some knowledge of elementary physics and chemistry will help in the study of anatomy and physiology and will also be needed to enable the nurse to care for the patient and to carry out nursing procedures.

Chemistry deals

1. with the composition of matter and the reactions between various types of matter
2. *Physics* deals with the behaviour and characteristics of matter, for example whether it gives off heat and light, or conducts electricity

MATTER

Matter is anything that can occupy space. When we refer to a 'space' we usually mean that it contains air, which can be displaced by other forms of matter. For example, an 'empty' bottle is not really empty as it contains air. A container from which the air has been extracted is called a *vacuum*.

Physical States of Matter

Matter has three physical states: (a) solid, (b) liquid and (c) gas.

A *solid* does not easily alter in either shape or size, e.g. stone, brick.

A *liquid* takes the shape of the vessel it is put into, but does not alter in size, e.g. 500 millilitres (ml) of water has no shape, but it takes the shape of the container it is put into; if it is put into a jug which will hold 1 litre the jug will not be filled.

A *gas* takes the shape and size of the vessel containing it. If the air is extracted from a container and a small quantity of gas is introduced the gas will expand to fill the container. If more gas is introduced the gas in the container is compressed, and the small particles which make up the gas are packed more closely together. More gas could be introduced until the flask bursts. A common example of this can be seen when a bicycle tyre is pumped up. After a few pumps the tyre seems full, but when someone gets on the bicycle the tyre goes flat because there is so little air in it. You can go on pumping until there is so much air in the tyre that when you get on the bicycle your weight does not flatten the tyre. A measured amount of gas in a cylinder will exert a known pressure. As the gas is used up the pressure in the cylinder will drop and it is in this way that the quantity of gas remaining in the cylinder can be gauged. Common examples of gases are oxygen, hydrogen and nitrogen.

Change of State

At normal temperatures iron is solid and water is liquid, but the state of matter can alter. The changes are due to heating or loss of heat. For example, liquid water becomes solid ice when heat is lost, and solid butter becomes liquid oil when it is warmed. The changes are described in Box 1.1.

Changes of state

- **Melting** is the change of a solid into a liquid as a result of heating, e.g. the turning of ice into water
- **Evaporation** is the change of a liquid into a gas as a result of heating, e.g. the turning of water into steam or water vapour. Gases formed by evaporation are called *vapours*
- **Condensation** is the turning of a gas into a liquid as a result of cooling, e.g. the turning of water vapour into water
- **Consolidation** is the turning of a liquid into a solid as a result of cooling, e.g. the turning of water into ice

Box 1.1

The first two of these changes, melting and evaporation, are due to heating, but the heat which causes this change of state does not cause any rise in the temperature of the matter. Heat is a form of energy and the energy in this case is used up in producing the change of state, so it does not heat the matter. For example, the temperature of melting ice is 0°C. When the ice melts the water is also at a temperature of 0°C. Water boils at 100°C. If you leave the kettle on the heat the water will not get any hotter because the added heat is used up in turning the water into water vapour. In the same way, heat from the body evaporates the sweat from the skin, excess body heat is used up and the body is cooled. Heat used up in this way is called *latent heat*. In condensation or consolidation latent heat is set free.

ELEMENTS, COMPOUNDS AND MIXTURES

Some substances cannot be split up and these substances are called *elements*. Over ninety elements are known to exist, including oxygen, carbon, nitrogen, iron, silver and gold. Substances which can be split into different elements are of two types: (a) compounds, and (b) mixtures.

A *compound* is a substance made of two or more elements which combine chemically to form a new substance with new properties. For example, hydrogen is a very light gas, used to fill balloons. It is highly inflammable and may explode. Oxygen is a gas which supports combustion but will not itself burn. These two gases in combination make a compound, water, which is a fluid, but it is not inflammable, will not explode, will not support combustion and will not burn. In fact it is very useful for putting out a fire. The elements which form a compound are always present in fixed proportions. For example, water consists of two parts of hydrogen combined with one part oxygen. If the proportions were altered a compound might be formed, but it would not be water. Two parts of hydrogen combined with two parts of oxygen give a compound called hydrogen peroxide which looks like water but has very different properties. Another characteristic of a compound is that it is not easy to separate the elements which compose it. To accomplish this separation a chemical change must take place.

A *mixture* is a substance made of two or more elements mixed together but not chemically combined. Air is a mixture of gases, mainly nitrogen and oxygen, with traces of carbon dioxide and other gases. A mixture has no new properties but simply those present in the elements which form the mixture. Air supports combustion because it contains free oxygen, but it does not support combustion as well as pure oxygen because of the high percentage of nitrogen, which will not support combustion. The elements in a mixture can be separated fairly easily because they are not chemically combined. Oxygen in air will support combustion because it is 'free' oxygen. Oxygen in water is not free and therefore will not support combustion. In a mixture there is no fixed proportion of elements composing it. Three different samples of air may contain three different proportions of oxygen, yet each is a specimen of air.

GAS PRESSURE

The molecules of a solid are strongly attracted to one another so a solid does not easily alter in size or shape. The molecules of a liquid are less strongly attracted to each other so their positions can be altered more easily and the liquid takes the shape of the container. The molecules of a gas are attracted to each other very little so that when gas is put into a vacuum container the molecules separate from one another and fill the flask. If more gas is introduced the molecules become more densely packed and the gas is said to be under pressure. The pressure of gases is important in their use in the body. Unless oxygen is under sufficient pressure in the lungs it will not be transferred to the blood for transport round the body and the body tissues will suffer from lack of oxygen. On a high mountain the *pressure* of the air is so low that man cannot live without an extra supply of oxygen, though a candle will burn because there is sufficient quantity of air even though the pressure is reduced. In an airless room where there is reduced quantity of oxygen the candle will not burn, though man can continue to live, as the oxygen present is still under pressure.

SOLUTIONS AND EMULSIONS

When a lump of sugar is put into a cup of hot water the sugar dissolves and a *solution* of sugar is present in the cup. The molecules of sugar mix with the molecules of water until the sugar is evenly distributed throughout the fluid. The process can be speeded up by stirring, but the even distribution of sugar will occur eventually without stirring. A measured quantity of fluid will only dissolve a measured quantity of another substance, and when the fluid has dissolved as much as possible it is called a *saturated solution*. Fluids can dissolve solids, liquids and gases. Oxygen is dissolved in water, which enables fish to live in it. Heating a fluid drives off the dissolved gas, which appears at the surface as bubbles as the fluid gets hot. Water is the most common solvent and can dissolve a wide variety of substances, but substances which cannot be dissolved in water may be dissolved in other liquids. For example, water will not dissolve oil, but oil will dissolve in methylated spirit. The fact that water is a common solvent makes it important in the body, two-thirds of which is water. Most of this water is in the cells which make up the body, some of it surrounds the cells, which can only live in a salt solution, called saline, and some is in the circulating blood and lymph.

An *emulsion* of oil is different from a solution. If oil is mixed with water the oil will rise to the top when stirring is stopped and will form a layer of pure oil on top of the water. If the oil is mixed with a solution of sodium carbonate (washing soda) an emulsion will be formed. The oil will be broken up into little droplets which will be suspended in the fluid and will not run together again. When allowed to stand the oil will still rise to the top, but the fat droplets will remain separated. Milk is a natural emulsion and the droplets of fat which form the cream can easily be seen under a microscope.

ACIDS AND ALKALIS

Substances may be either acid or alkaline in reaction, or they may be neutral (neither acid nor alkaline). The simplest agent used to detect the reaction of a substance is *litmus* which can be obtained as a fluid, but is conveniently used in the form of litmus paper. Blue litmus paper will turn pink when in contact with acid. Pink litmus paper will turn blue when in contact with an alkali. Neutral substances turn both pink and blue litmus paper a purple colour. Every living cell must have the correct level of acidity or alkalinity if it is to survive. The blood and body tissues are slightly alkaline and vary little throughout life. If too much acid or too much alkali is present in the body illness will occur, and if there is an alteration in the reaction of the blood or body tissues death will follow. It is necessary to know the degree of acidity or alkalinity of certain substances and for this purpose a universal indicator is used. It changes colour from bright red, indicating a strong acid, to purple, which indicates strong alkali. These colour changes are spoken of as the pH of the substance being tested, and the scale runs from pH 1 for the strong acid to pH 14 for the strong alkali. Neutral is pH 7 and the pH of the blood is 7.4.

Pure water is neutral in reaction and as previously stated consists of molecules composed of two atoms of hydrogen and one atom of oxygen. Some of the molecules are broken up into smaller particles, e.g. hydrogen ions (H ions) or hydroxyl ions (OH ions). H ions have a positive electrical charge because the atoms have lost an electron and OH ions a negative electrical charge. In pure water the H ions and the OH ions are equal in number and the reaction is neutral. In some solutions H ions may outnumber OH ions and the reaction of the solution will be acid. If the OH ions outnumber the H ions the reaction will be alkaline. The pH shows the concentration of hydrogen ions (the *H ion concentration*), strong acids, e.g. hydrochloric acid, having a high H ion concentration, and strong alkalis, e.g. caustic soda, the highest concentration of hydroxyl ions (OH ions). Some substances, which slow down changes in pH, are called *buffer substances*. In an alkaline solution such as body fluids, buffer substances neutralize acids or any excess of alkalis which may be added to it or produced in it. Examples of buffer substances are sodium, potassium and protein which serve as buffers against acid substances in the body and carbonic acid, lactic acid and fatty acids which serve as buffers against any excess of alkaline bases.

MINERAL SALTS

A mineral salt is a substance made by the action of acid on a mineral. The mineral is called the *base* of the salt, and the scientific name shows the base and the acid which together form the salt, e.g. sodium chloride is made by the action of hydrochloric acid on sodium. A salt may be acid, alkaline or neutral in reaction. A strong acid acting on a strong alkali forms a neutral salt; a strong acid acting on a weak alkali forms an acid salt; a weak acid acting on a strong alkali forms an alkaline salt. The various salts present in the body differ only slightly in reaction.

MEASURING THE AMOUNT OF A SUBSTANCE

The number of neutrons added to the number of protons in an atom gives its *atomic weight*. Hydrogen has an atomic weight of 1 (as it has one proton and no neutrons) and oxygen an atomic weight of 16 (as it has eight protons and eight neutrons). A molecule of water, which consists of two hydrogen atoms and one oxygen atom (H_2O), will therefore have a weight of 18. This is known as its *molecular weight*. Understanding atomic and molecular weights is important in understanding how the amount of a substance is measured. The concept of atomic weight will be explained in more detail in Chapter 2.

The amount of a substance can be expressed in terms of its mass, but in physiology and biochemistry it is more important to know *how many particles* (atomic, ionic or molecular) of the substance are present rather than their total mass. However, since even tiny amounts of a substance will contain many millions of particles, it would be very cumbersome to use such huge numbers. Therefore a unit defined as a standard large number of particles was devised. The number of particles chosen was the number of particles in that amount of a substance that has a mass equal to its molecular or atomic weight in grams; this number will always be $6 \cdot 023 \times 10^{23}$ and is referred to as Avogadro's number. For example, the atomic weight of carbon is 12, therefore 12 g of carbon will contain $6 \cdot 023 \times 10^{23}$ atoms of carbon. The molecular weight of water is 18, therefore 18 g of water will also contain $6 \cdot 023 \times 10^{23}$ molecules of water. The amount of substance containing this number of particles is called a *mole*, the symbol for which is mol.

The mole can be used to indicate the concentration of a substance in a solution. If one mole of a substance is added to one litre of water, the concentration of the substance in the resulting solution will be 1 mol/l. Because the mole is a measure of the number of particles, one mole of a molecule which breaks up into ions when added to water will produce one mole of each of the ions produced. For example, hydrochloric acid (HCl) breaks up into hydrogen ions (H^+) and chloride ions (Cl^-) when added to water. One molecule of HCl will produce one H^+ and one Cl^-. Therefore $6 \cdot 023 \times 10^{23}$ molecules of HCl (one mole of HCl) will produce $6 \cdot 023 \times 10^{23}$ H^+ (one mole of H^+) and $6 \cdot 023 \times 10^{23}$ Cl^- (one mole of Cl^-).

The mole is often too large a unit to describe the concentrations of substances encountered in clinical medicine. Therefore a prefix is used, as for other units. The prefix commonly used is milli-, one millimole (mmol) being 1000 times smaller than a mole. For example, the concentration of sodium chloride (NaCl) in the blood is approximately 0·15 mol/l, but this is usually written as 150 mmol/l. A solution made up to have this same concentration of NaCl (i.e. 150 mmol/l) is known as *isotonic* or *physiological saline*.

Moles cannot be used for measuring the concentration of protein substances because the molecules are so complex that their molecular weight cannot be determined accurately. Such substances are therefore expressed in terms of their mass, i.e. in grams per litre or in milligrams (mg) per 100 ml. It is rarely necessary to convert from one to the other but if it should be the formula is:

$$\text{Concentration in mmol/l} = \frac{\text{Concentration in mg/100 ml} \times 10}{\text{Molecular weight of substance}}$$

The concentrations of dissolved gases, e.g. in the blood, are often stated in terms of pressure rather than moles or grams because pressures are easier to measure.

DIFFUSION

If two gases of different composition come into contact, intermingling of the gases takes place until the composition of both is the same. For example, the air we breathe out contains more carbon dioxide and less oxygen than the air we breathe in, but the expired air does not float around the room as a portion of different air. The air in the whole room will eventually contain less oxygen and more carbon dioxide overall. This process is known as *diffusion* (Fig. 1.1).

Fig. 1.1   Diagram to show the diffusion of liquids.

The term diffusion is also used to describe the passage of small molecules of acids and salts through a semipermeable membrane. A membrane may be described as permeable when it allows substances in solution to pass through; as impermeable when no fluid can pass through; and as semipermeable when water and small molecules in solution can pass through but not larger molecules. If a strong solution of salt is separated from a weaker solution by a semipermeable membrane the salt molecules will pass through the membrane from the strong solution to the weaker one until both solutions are of equal strength.

OSMOSIS

If a substance with large molecules, e.g. sugar, is made into a solution and is separated from a weaker sugar solution by a semipermeable membrane only water will pass through the membrane, from the weak solution to the strong solution, because the molecules are too large to pass. This passage of water across a semipermeable membrane is called *osmosis* (Fig. 1.2).

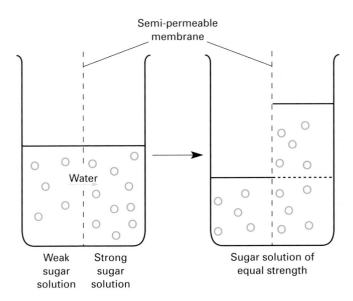

Fig. 1.2   Diagram to show osmosis.

SPECIFIC GRAVITY

Specific gravity is the weight of a known volume of liquid divided by the weight of an equal volume of pure water.

$$\text{Specific gravity} = \frac{\text{Weight of substance}}{\text{Weight of equal volume of water}}$$

The specific gravity of pure water is expressed as 1·000. The specific gravity of urine ranges between 1·010 and 1·020, and that of blood is about 1·055.

ORGANIC MATTER AND INORGANIC MATTER

Organic matter is that which is living or has been alive, e.g. wood or coal. Inorganic matter is that which is not living and never has been alive, e.g. water or iron. Inorganic substances can be built up into organic compounds by living organisms.

All living matter is made up of small units called *cells*. Some organisms, such as bacteria, consist of one single cell; others, like man, consist of many hundreds of millions of cells, all functioning together to make a complete whole. All living cells, however simple, have certain characteristics which are always present. These characteristics are listed in Box 1.2.

Characteristics of living matter

- Activity
- Respiration
- Digestion and absorption of food
- Excretion
- Growth and repair
- Reproduction
- Irritability

Box 1.2

Activity

This is the most striking characteristic of living matter. It is more apparent in the animal world than in the vegetable world, as the animal must move about in search of food, but even in plants the buds can be seen to break out in the spring, and with a microscope activity becomes as obvious in the plant as in the animal. There can never be any activity without energy. Cars are driven by energy derived from petrol, trains are generally moved by electrical energy, and in living matter the energy is also obtained by burning fuel. The fuel for the human body is the food eaten, particularly carbohydrates and fats. Oxygen is also necessary for the combustion of fuel, and living things obtain their oxygen from the air, or water, in which they live. Combustion of fuel also produces waste products, such as carbon dioxide and water, which must be disposed of.

The combustion of fuel in living matter produces some energy for work and some energy in the form of heat. The body is economical in that it produces more work energy and less heat for each unit of food consumed, but the heat produced by combustion is not altogether wasted, for some heat is needed. The body can remain healthy only within a narrow temperature range of 36–37·5°C and any heat produced in excess of this must be got rid of if life is to continue.

Respiration

All living matter requires oxygen and gives off carbon dioxide. The oxygen is required for combustion, or oxidation, of food, and carbon dioxide is the waste product of combustion. This process of taking in oxygen and giving off carbon dioxide is called *respiration* and it is continuous throughout life. The amount of oxygen required and the quantity of carbon dioxide given off vary with the amount of activity taking place. During sleep the human body requires comparatively little oxygen, but it requires much more during strenuous exercise such as playing football or climbing a mountain.

Digestion and Absorption of Food

All living matter requires food. Some food can be absorbed as it is, but most of it must be broken down into substances which have smaller and more simple molecules before it can be absorbed. The process of breaking down complex foods into smaller substances is termed *digestion*. It is brought about by enzymes, which are themselves protein substances and which act upon the food, preparing it for *absorption*.

Excretion

All living matter produces waste products which are of no further use and must be got rid of. If allowed to accumulate, waste products will interfere with the processes of life. In the human body these waste products include carbon dioxide from the lungs, sweat from the skin and urine from the kidneys.

Growth and Repair

Living matter is able to build up new living matter, similar to itself, from food which is taken in. Protein foods include meat, cheese and milk, and these provide the building material for the body. Because the body is continually active its parts are constantly being worn out and they are made good by the building of new living matter. In childhood, growth and repair are taking place simultaneously. In old age the building up does not keep pace with the breaking down, so weakness begins to appear and will eventually cause death unless illness supervenes. It is this ability to grow and to repair its own tissues, and also the power of reproduction, that makes living matter different from man-made substances.

Reproduction

All living matter can reproduce its own kind. In the simplest forms of life reproduction is a very simple process, involving the splitting of the parent cell into two. In animals, female cells called *ova* are produced in the ovaries, and *sperms*, or male cells, are produced in the testes. The sperm cell must fertilize the ovum before reproduction can take place.

Irritability

All living matter is able to respond to a stimulus, and in higher animals this means the power to be aware of the environment and to respond to it. Irritability is more marked in animals, but it can be observed in plants. If a bulb is planted upside down the roots will still grow deep into the soil and the leaves into the air, or if a plant is placed in a dark corner it will grow tall and thin as it seeks the light. This indicates that even plants are aware of the environment. In animals, irritability allows the awareness of danger, the recognition of that which is useful, e.g. food and water, and the exercise of free will.

All this is what is meant when the physiologist talks about 'life'. Living matter is that which has all the characteristics mentioned above, and all these characteristics are the result of chemical changes. All living matter has the ability to produce protein substances called *enzymes* which in turn control and produce chemical changes within the cell. Enzymes cause other substances to combine with one another, or to split up, though they themselves do not enter into the reaction. For example, complex starches combine with water and split up into simple sugars in digestion; in metabolism, sugar splits up into carbon dioxide and water. All these changes are brought about by enzymes.

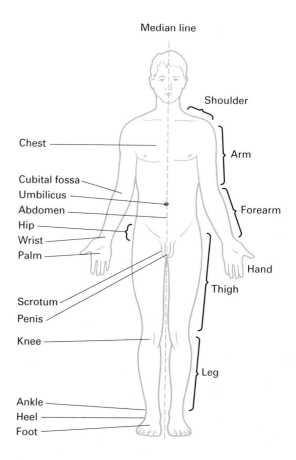

Fig. 1.3 The anatomical parts of the male body (front view).

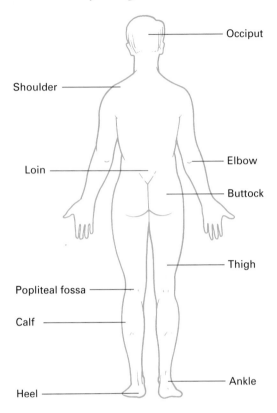

Fig. 1.4   The anatomical parts of the body (back view).

DEFINITION OF TERMS USED IN ANATOMY

In order to achieve uniformity of description an *anatomical position* has been chosen and defined. The body is erect, facing the observer, with the arms at the sides and the hands facing forwards (Figs 1.3 and 1.4). Terms that are commonly used are given in Box 1.3.

The *median* or *sagittal line* is an imaginary line passing vertically through the midline of the body from the crown of the head to the ground between the feet, dividing it into right and left halves. *Lateral* means furthest from the median line; *medial* means nearest to the median line.

A *horizontal* section divides the body into superior and inferior portions.

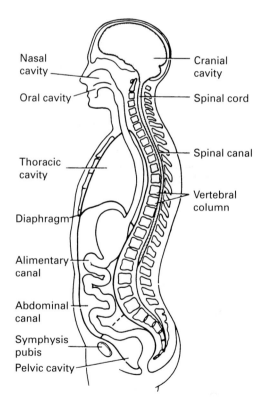

Nasal cavity

Oral cavity

Thoracic cavity

Diaphragm

Alimentary canal

Abdominal canal

Symphysis pubis

Pelvic cavity

Cranial cavity

Spinal cord

Spinal canal

Vertebral column

Fig. 1.5 Diagrammatic section of the cavities of the head and trunk (the unlabelled broken line marks the boundary between the abdomen and the pelvic cavity).

A *sagittal* section divides the body into right and left portions parallel to the median line.

A *coronal* section divides the body into anterior and posterior portions.

Cavities of the Body

The body has two major cavities, each subdivided into lesser cavities (Fig. 1.5).

The *ventral cavity*, within the trunk, is subdivided into:

1. The thorax or thoracic cavity
2. The abdomen, or abdominal cavity which is continuous with the pelvic cavity

The *dorsal cavity* is subdivided into:

1. The cranial portion, containing the brain
2. The spinal portion, containing the spinal cord

SUMMARY

This chapter has reviewed some of the physical properties of matter and also outlined the anatomical organization of the body. The remainder of this section is concerned with physiological organization of the body into cells, tissues and organs, with a brief consideration of homeostasis. Further anatomical details will be given in the following sections which look at the systems of the body.

Chapter 2
The Chemical Level of Organization

This chapter covers some basic concepts in chemistry, including chemical bonding, and applies this to the range of biological macromolecules—fats, carbohydrates, proteins and nucleic acids.

THE STRUCTURE OF MATTER

All matter is made up of tiny particles called *molecules* which are so small that they cannot be seen with an ordinary microscope. Molecules are held together only by the attraction of one molecule to another, in the same way that a needle is held to a magnet. It is perhaps rather difficult to accept the fact that an iron bar is made of molecules which do not even adhere to each other but are only held together by mutual attraction. It is, however, an important concept which must be remembered. The force of attraction is called *cohesion* and must be distinguished from adhesion. A molecule is the smallest particle which can exist alone but which still has all the properties of the substance. For example, a molecule of water has all the properties of water which were mentioned earlier.

A molecule can be divided into smaller particles called *atoms*, but an atom cannot exist alone. A molecule of an element consists of atoms of that element, e.g. a molecule of oxygen consists of two atoms of oxygen (written O_2). A molecule of a compound consists of atoms of the different elements present in the compound, e.g. a molecule of water consists of two atoms of hydrogen and one atom of oxygen (written H_2O). A molecule of carbon dioxide consists of one atom of carbon and two atoms of oxygen (CO_2) and a molecule of glucose, which is a simple form of sugar, consists of six atoms of carbon, twelve atoms of hydrogen and six atoms of oxygen ($C_6H_{12}O_6$). These are examples of very simple chemical compounds. In the more complicated compounds which form the human body many elements may be present, and the number of atoms of any one element may run into hundreds. Although all molecules are very small they vary considerably in size and complexity.

Atomic Number and Atomic Weight

The word 'atom' means indivisible and the name was given when scientists thought that the atom could not be divided. Now, however, more is known about the structure of an atom and most people have heard about 'splitting the atom' and its connection with atomic energy. An atom consists of even smaller particles of three different types: *protons* which carry a positive electrical charge, *electrons* which carry a negative electrical charge and *neutrons* which are neutral. The protons and neutrons are massed together to form the nucleus of the atom, while the electrons are arranged around the nucleus in one or more energy levels. Different elements have different numbers of protons and electrons forming their atoms and the number of protons is equal to the number of electrons. This number is called the *atomic number*. Hydrogen has one proton in its nucleus and one electron circling round it and its atomic number is one. It has no neutron. Carbon has six protons and six neutrons in its nucleus and six electrons, arranged round it, so its atomic number will be six. Sodium has eleven protons and twelve neutrons in the nucleus, and eleven electrons round it, so the atomic number is eleven (Fig. 2.1). These numbers do not have to be memorized because they can be obtained from tables. The presence of neutrons in the nucleus does not affect the atomic number.

The weight of a neutron is roughly equal to the weight of a proton and the number of neutrons added to the number of protons gives the *atomic weight*. The weight of an electron is so small that it is not significant. Hydrogen has an atomic weight of one as it has one proton and one electron but no neutron. Carbon has an atomic weight of twelve as it has six protons and six neutrons in its nucleus. Sodium has an atomic weight of twenty-three as it has eleven protons and twelve neutrons in its nucleus.

ORGANIC CHEMISTRY

There are rules governing the distribution of electrons between energy levels and these rules help us to understand *chemical bonding*. The maximum number of electrons in the first energy level is two and the maximum number in the second energy level is eight. The electrons are usually represented as circling the nucleus in two orbits. The electrons in the second energy levels, however, are

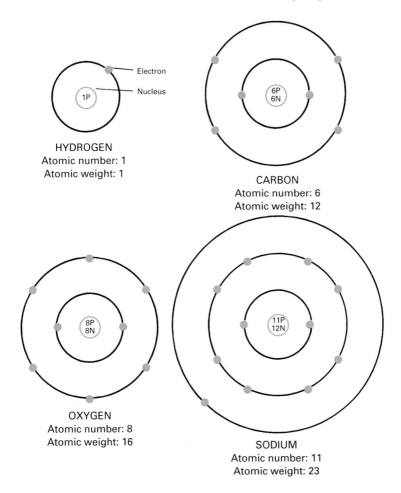

Fig. 2.1 Diagram of the structure of atoms of hydrogen, carbon, oxygen and sodium. P, proton; N, neutron.

arranged in four *orbitals* which project from the nucleus in such a way as to form the four corners of a tetrahedron (Fig. 2.2).

The rules governing the distribution of electrons between orbitals is that each orbital must have at least one electron and can only have a maximum of two. If these simple rules are applied to carbon then there is a distribution of one *unpaired electron* in each second energy level orbital. Oxygen, with six electrons in the second energy level, has two paired electrons and two unpaired electrons. The number of unpaired electrons is important as this is equal to the *valency* of the element and this dictates how many

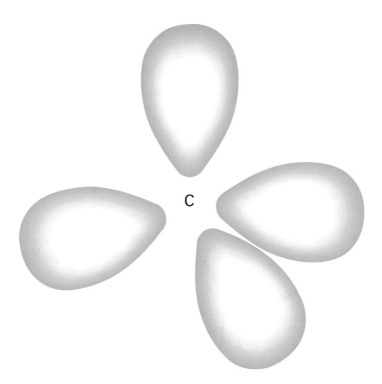

Fig. 2.2 Diagram of the four second-energy-level orbitals of a carbon atom.

bonds the element, in its atomic form, can make with other elements. Hydrogen, due to its single unpaired electron in the first energy level, has a valency of one.

Covalent Bonding

Elements with unpaired electrons can share these with other similar elements and thereby form *covalent bonds*. One atom of carbon can form covalent bonds with four atoms of hydrogen to form the simplest organic *hydrocarbon* compound, methane (CH_4). Oxygen can bond covalently with carbon and hydrogen to form acetone (CH_3COCH_3). Another compound containing carbon, hydrogen and oxygen is ethanol (C_2H_5OH) (Fig. 2.3). Covalent bonds are relatively strong bonds requiring energy, for example in the form of heat, to break them. Once they are broken they release energy and this is why we burn fossil fuels such as coal and oil. It also

```
        H
        |
  H —  C  — H
        |
        H
```
Methane (CH_4)

```
        H    O    H
        |    ||   |
  H —  C  —  C  —  C  — H
        |         |
        H         H
```
Acetone ($CH_3CO\ CH_3$)

```
        H    H
        |    |
  H —  C  —  C  — OH
        |    |
        H    H
```
Ethanol (C_2H_5OH)

Fig. 2.3   Some simple organic compounds which illustrate the principles of covalent bonding.

explains the energy content of foods such as fats and carbohydrates. Nitrogen, which has a valency of three, also forms organic compounds and is incorporated in the amino and nucleic acids which will be described below.

Ions and Electrolytes

Atoms do not remain static as electrons may be split off from one atom and gained by another. Atoms are constantly gaining and losing electrons. When a substance is in solution in a fluid and an electron splits from an atom of the substance, an *ion* is formed. Substances which dissociate in this way are known as *electrolytes* because they carry small electric charges. Electrolytes are present in body fluids and are essential to life.

Sodium and chlorine are examples of electrolytes. Sodium readily loses the unpaired electron in its outer energy level giving it a positive charge, and chlorine can accept an electron into its outer energy level giving it a negative charge. When sodium and chlorine form a compound, the common salt sodium chloride (NaCl), the bond that exists between them is called an *ionic bond*. These bonds

Fig. 2.4   Diagrammatic representation of hydrogen bonding in water. The broken lines are the weak interactions between the oxygen and hydrogen atoms.

are easier to break than covalent bonds and this is demonstrated by the fact that salt can be dissolved by adding it to water.

Hydrogen Bonds

The best example of *hydrogen bonding* is in water (Fig. 2.4) where there is an unequal distribution of the electrons in the covalent bond between the hydrogen and the oxygen. The electrons are more attracted to the oxygen giving it a slight negative charge and the hydrogen a slight positive charge. Such bonding means that water is a liquid and not a vapour.

There are other examples of hydrogen bonds. These are very weak bonds but they occur in such abundance in some organic compounds, particularly in the biological macromolecules such as proteins and nucleic acids, that they contribute greatly towards the stability of these compounds.

BIOLOGICAL MACROMOLECULES

These include the triglycerides (fats), polysaccharides, proteins and nucleic acids which have important nutritive, structural and genetic functions (Box 2.1). All these molecules are *polymers*, which means that they are built from covalently bound repeating units. Fats are, essentially, long chain hydrocarbon molecules attached to glycerol. *Polysaccharides* are built from common carbohydrates (sugars) such as glucose; *proteins* are built from amino acids; and *nucleic acids* are built from purine and pyrimidine *bases*.

Biological macromolecules

- Triglycerides (fats)
- Carbohydrates
- Proteins
- Nucleic acids

Box 2.1

Each of these polymers has its own type of covalent bond. The hydrocarbon chains of fats are attached to glycerol by *ester linkages*. The polysaccharides have *glycosidic linkages*, the proteins have *peptide bonds* and the nucleic acids have *phosphodiester bridges*. One of the commonest polysaccharides is *starch*. All of the enzymes are proteins and DNA (deoxyribonucleic acid), the material from which our genes are made, is a nucleic acid.

Primary, Secondary, Tertiary and Quaternary Structure

The biological macromolecules display various levels of structure and, using protein as an example, these can be explained as follows (Box 2.2). The sequence of amino acids, of which there are twenty in organic compounds, is described as the *primary structure*.

Structural aspects of biological macromolecules

- Primary
- Secondary
- Tertiary
- Quaternary

Box 2.2

Specific regions of this primary structure which are regular in structure due to folding into helices and sheets are described as the *secondary structure*, and any subsequent folding is described as the

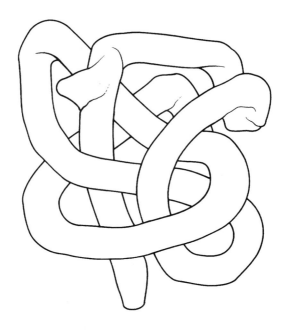

Fig. 2.5   The tertiary (folded) structure of a globular protein.

tertiary structure (Fig. 2.5). Where two or more molecules are held together this is described as the *quaternary structure* (Fig. 2.6). The primary structure is held together by covalent bonds, and a major factor in the higher order of structure is hydrogen bonding between regions of primary structure, although other forces, including covalent bonding, can maintain these structures.

THE CENTRAL DOGMA OF GENETICS

DNA consists of two strands held together in a *double helix* by hydrogen bonds (Fig. 2.6). The primary structure of the bases in DNA is the basis of the *genetic code* which gives rise to all of our characteristics. The code consists of triplets (known as *codons*) each of which leads to the incorporation of an amino acid into a protein molecule.

DNA can *replicate* and conserve the genetic code during cell division (see Chapter 3) and can also be *transcribed* into RNA

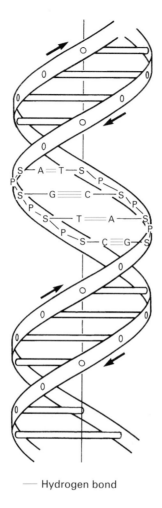

—— Hydrogen bond

Fig. 2.6 The quarternary structure of DNA is maintained by hydrogen bonds between the two strands.

(ribonucleic acid) (Fig. 2.7). RNA also contains the genetic code and is used as a template for the building of unique protein molecules. This process is, obviously, under very close regulation in living cells in order to ensure that the right protein molecules are made as required. Disturbance of this process and the unregulated production of particular cells is called 'cancer'.

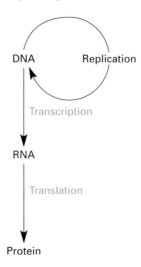

Fig. 2.7   The central dogma of genetics. Replication and transcription take place in the nucleus. Translation takes place in the cytoplasm.

SUMMARY

The main points to be appreciated from this chapter are that biological macromolecules are covalently bonded polymers. The more complex polymers—proteins and nucleic acids—also show weak interactions such as hydrogen bonding between chains.

Chapter 3
The Cellular Level of Organization

The cell is the unit of living matter, the fundamental building block of life, and all cells have certain features in common. This chapter will consider those features and highlight the structure and function of the limiting membrane around the cell and the processes whereby cell division takes place.

STRUCTURE OF A CELL

All cells are made of a substance called *protoplasm*, which is jelly-like, opaque and colourless and consists mainly of water with other substances in solution (Fig. 3.1).

The word *cytoplasm* is commonly used to describe the proto-plasm which forms the bulk of the cell, the prefix 'cyto' being derived from the Greek word for cell. The cytoplasm contains

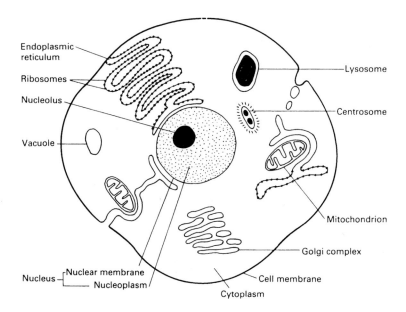

Fig. 3.1 Diagram of a cell.

Phospholipid
bilayer

Membrane protein molecules

Fig. 3.2 Diagram of the plasma membrane.

molecules of ribonucleic acid (RNA) which acts as a messenger carrying information out from the nucleus to the cytoplasm.

The *plasma membrane* plays a very important role in holding the contents of the cell together, and yet is flexible enough to allow cells to conform to many shapes. The plasma membrane also regulates the passage of materials such as nutrients, electrolytes and waste products in and out of the cell. In this way the plasma membrane is described as being *selectively permeable* or *semipermeable*.

The plasma membrane (Fig. 3.2) is composed of protein and a substance called *phospholipid*. The structure of phospholipid molecules which makes them *hydrophobic* (repelled by water) at one end and *hydrophilic* (attracted to water) at the other end leads to the unique structure of the plasma membrane, a double layer known as a *phospholipid bilayer*. The protein molecules are embedded in this bilayer and it is the protein molecules which confer the selective permeability on the plasma membrane. The protein molecules can form channels or carriers for the transport of certain substances across the plasma membrane.

Other functions of the protein in the plasma membrane are to act as sites of recognition between cells and to act as receptors whereby substances such as hormones (see Chapter 8) can attach to cells and influence their ability.

TRANSPORT PROCESSES ACROSS PLASMA MEMBRANES

Movement of substances across cell membranes takes place by several processes (Box 3.1). The processes of *osmosis* and *diffusion*, whereby water and substances in solution, respectively, move across semipermeable membranes, have already been described in Chapter 1, and these work across the plasma membrane.

Movement across the plasma membrane

- Osmosis
- Passive diffusion
- Facilitated diffusion
- Active transport

Box 3.1

The molecules of some substances are unable to diffuse across the cell membrane. The reasons for this are that the molecules may be too large or may be repelled by the hydrophobic lipid bilayer. One of the processes by which such substances, which the cell requires, can cross the plasma membrane is *facilitated diffusion*. Facilitated diffusion can only take place where the concentration of the substance outside the cell is greater than the concentration inside the cell. However, there are specific sites in the plasma membrane (protein carriers) at which the substances can cross. Some cells accumulate nutrients such as carbohydrates in this way.

A further process whereby substances can be accumulated or expelled from cells is called *active transport*. This process requires specific carriers in the membrane, but differs from facilitated transport in that it can accumulate substances regardless of their concentration on either side of the plasma membrane, and it achieves this by using energy. A good example is the way in which cells accumulate potassium (K^+) and expel sodium (Na^+) ions and thereby maintain a high concentration of potassium inside the cell relative to the outside. The cells of the liver accumulate glucose in the same way.

The nucleus is a dense body within the cell. It contains nucleoplasm and is surrounded by a *nuclear membrane*. The characteristic compounds of the nucleus are deoxyribonucleic acids (DNA), which contain the genetically inherited information required for the maintenance of the cell. The nucleoplasm stores the information necessary for the cell to grow and to divide into two daughter cells. This information is stored in the *genes*, which are folded up to form *chromosomes*.

Chromosomes are normally only visible under the microscope when the cell is about to divide.

The *mitochondria* are the power stations of the cell. They are responsible for converting the food ingested by the cell into energy.

Vacuoles are clear spaces within the cytoplasm, which contain waste materials, or secretions, formed by the cytoplasm.

The *centrosome* is a small rod-shaped body near the nucleus and is important in cell division. It is surrounded by a radiating thread-like structure and contains two centrioles.

CELL REPRODUCTION

Mitosis

In the human body, cell division takes place by a process called *mitosis* (Fig. 3.3). There are seven stages (Box 3.2):

1.   The centrosome divides into two and each moves away from the other, though they are still attached by thread-like structures. This stage is called the *prophase*.
2.   The nuclear material forms dark thread-like structures which are the chromosomes. There are forty-six chromosomes in human cells, though the number differs in other species.
3.   The nuclear membrane disappears and the chromosomes arrange themselves around the centre of the cell. They appear to be attached to the thread-like structure of the centrosomes, which are by now at either end of the cell. These two changes are known as the *metaphase*.
4.   The chromosomes divide longitudinally into two equal parts.
5.   The two groups of chromosomes move away to either end of the cell and arrange themselves around the centrosomes. The thread-like structures joining the centrosomes now divide. These two changes are called the *anaphase*.
6   The cell body gets narrower round the centre. The thread-like structures disappear and two nuclear membranes reappear. This is called the *telophase*.
7.   The cell divides and the chromosomes disappear into the nucleus. The daughter cells will in turn grow and will reproduce by mitosis.

The cells produced by mitosis are called *somatic cells*.

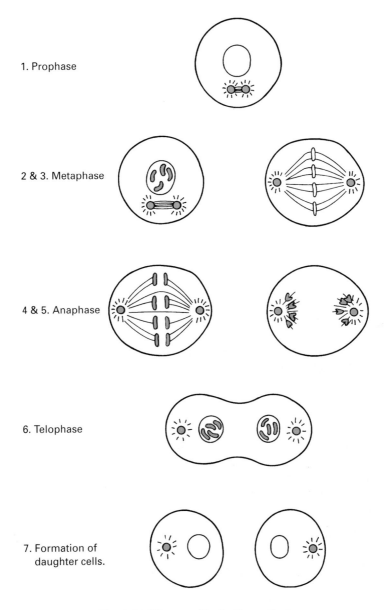

1. Prophase

2 & 3. Metaphase

4 & 5. Anaphase

6. Telophase

7. Formation of daughter cells.

Fig. 3.3 Diagram illustrating mitosis.

Stages of mitosis

- Interphase
- Prophase
- Metaphase
- Anaphase
- Telophase
- Interphase

Box 3.2

Meiosis

Reproduction in higher animals, including man, depends on the fusion of a spermatozöon from the male parent and an ovum from the female. These reproductive cells are also called *gametes*. Each gamete must receive half as many chromosomes as somatic cells so that when they unite in fertilization the full complement of chromosomes is obtained. As the sex cells mature two processes of cell division occur; the first is mitosis, in which each daughter cell receives a complete set of chromosomes. Following this a two-stage cell division occurs which is peculiar to reproductive tissue and is called *meiosis* (Fig. 3.4). The first division is similar to mitosis and gives rise to two daughter cells each containing the original number of chromosomes; the second division follows the first very quickly and results in four gametes each containing half the number of chromosomes. In man normal body cells have forty-six chromosomes (in twenty-three pairs) while each gamete has only twenty-three single chromosomes. Following fusion of the gametes the resulting cell, called a *zygote*, has forty-six chromosomes (in twenty-three pairs). Cell division of a zygote is by mitosis and the resulting multicelled organism is called an *embryo*.

The chromosomes consist of a chain of genetic material (DNA), and it is the genes which pass on the characteristics of the parent cell, so that daughter cells are always similar to parent cells. In this way some characteristics of children are derived from parents, including such characteristics as hair colour, height, degree of intelligence and many others. In any pair of genes one will exert a stronger influence than the other. The stronger gene is termed *dominant* and the weaker one *recessive*. The characteristics which are hereditary depend upon the dominance of the genes.

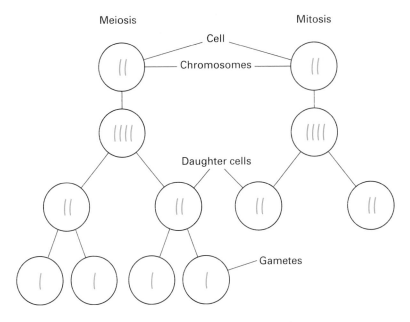

Fig. 3.4 Diagram comparing meiosis and mitosis.

Sex Determination

One pair of chromosomes from the father and one pair from the mother are the sex chromosomes which will determine the sex of the child. In the female the sex chromosomes are the same and are called XX. In the male they are different and are called XY. One chromosome from each pair will determine the sex of the child. If the child has an X chromosome from the mother and an X chromosome from the father it will be a girl (XX). If the child has an X chromosome from the mother and a Y chromosome from the father it will be a boy (XY) (Fig. 3.5).

EUKARYOTES, PROKARYOTES AND AKARYOTES

The cells of which the human body and all higher organisms, such as animals and plants, are comprised are called *eukaryotic cells*. The cells of *bacteria* are different in some respects and these are called *prokaryotic cells*. Some of the differences between eukaryotic and prokaryotic cells are that the cell membrane in prokaryotes is not a phospholipid bilayer and that the prokaryotic cell has no nucleus. Prokaryotic cells divide by a simple process of *binary fission*.

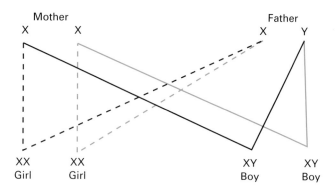

Fig. 3.5   Diagram to show sex determination.

Akaryotes have no cellular structure but do contain protein and genetic material (either DNA or RNA). *Viruses* are the best example of akaryotes.

MULTICELLULAR ORGANISMS

Multicellular organisms consist of many cells. Each cell is living and requires food, oxygen, water, suitable temperature and correct pH, but it is dependent on other cells to supply it with the necessities of life in return for the specialized work it carries out. For example, the cells of the lungs take in oxygen, and the cells of the digestive tract absorb food. Each cell develops in a specialized manner so that it can carry out its task satisfactorily. This specialized development is known as the *differentiation* of cells. Groups of specialized cells, developed for a particular purpose, form the tissues of the body, e.g. muscle tissue for movement and bone for support.

The multicellular organism in which we are particularly interested is the human body, a complex arrangement of millions of cells, but still having the basic characteristics of all living matter.

Chapter 4
Tissues, Organs, Systems and Homeostasis

There is a logic to the way in which the body is organized anatomically and physiologically, and this can best be studied by looking at organizational levels in the body above that of the cellular level. This chapter begins with the next level of organization, that of tissues, where more than one cell comes together to perform a similar function, up to the level of systems. Finally, a major feature of physiology, that of homeostasis, will be briefly considered.

TISSUES

The body consists of countless cells developed to form various different types of tissue. Tissue originates from a single typical cell, the egg cell or ovum, which is composed of protoplasm and contains a nucleus. After fertilization this cell multiplies and, by the process of differentiation, develops into all the various tissues required to form the different organs and parts of the body.

In the very early stages the ball of cells is divided into three layers. The outer layer is called the *ectoderm*, and from it the outer part of the skin develops with its nails, hair follicles and sweat glands and other epithelial tissues including the mucous membrane lining the nose and the mouth and the enamel covering the teeth. The nervous system also originates in the ectoderm. The middle layer is called the *mesoderm* and from it develop muscle, bone, fat and some of the internal organs, including parts of the cardiovascular system. The inner layer is called the *entoderm* and from it the lining of most of the alimentary and respiratory tracts develop.

A *tissue* consists of cells and the products of cells specially developed for the carrying out of a special function. In the body there are four main types of tissue (Table 4.1, Box 4.1).

Table 4.1   Classification of Tissues

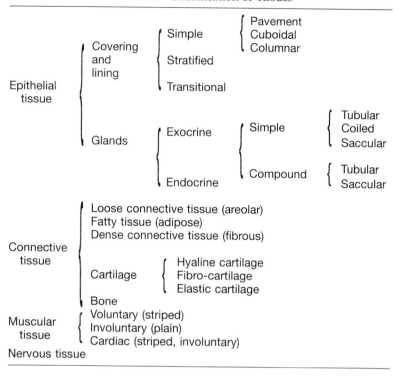

Main types of tissue

- Epithelial tissue or epithelium
- Connective tissue
- Muscular tissue
- Nervous tissue

Box 4.1

Epithelial Tissue

Epithelial tissue provides covering and lining membranes for the free surfaces inside and outside the body and is the tissue from which the glands of the body are developed. Epithelia protect underlying tissue from wear and tear, but must continually be renewed as needed. Some cells are specially developed to be able to absorb substances; these linings are only one cell in thickness and often have a specialized surface called a 'brush border'. Some epithelial tissue, particularly glandular tissue, has the ability to secrete substances manufactured within the tissue. Epithelia do not contain blood vessels, the nearest being in the underlying connective tissue, which may be some distance away. The cells are arranged on a 'basement membrane' which plays a part in binding them together.

Covering and Lining Epithelia
Covering epithelia may be classified according to the arrangement and shape of their cells.

Simple epithelium is composed of a single layer of cells attached to a basement membrane; it is very delicate and is found where there is little wear and tear (Fig. 4.1).

1. *Simple pavement epithelium* is composed of flat cells, which form a smooth lining. These may be found lining the blood vessels and forming the peritoneum.
2. *Simple cuboidal epithelium* has cells shaped like cubes and is found covering the ovary.
3. *Simple columnar epithelium* is composed of taller cells, packed onto a basement membrane. It is found where wear and tear is a little greater, for example. lining the stomach and intestines. According to the functions performed columnar epithelium may be modified.

Ciliated columnar epithelium has microscopic hair-like processes

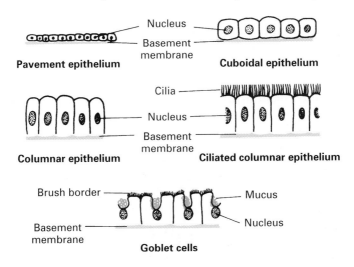

Fig. 4.1   Diagram showing types of simple epithelium.

projecting from the free surface of the cell. The cilia move together with a wave-like motion and mucus and other particles are moved along. This type of tissue is found in the respiratory tract.

Goblet cells are cells which secrete mucus, the mucus collecting in the cells until the cytoplasm becomes distended.

A *brush border* is found on cells specialized for absorption. These have minute finger-like projections which increase the area through which absorption can occur. They are found in the small intestine.

Stratified epithelium is made up of many layers of cells (Fig. 4.2). The deepest cells, called the *germinal layer*, lie on the basement membrane and are columnar. As they divide, and this occurs frequently, the parent cells are pushed nearer the surface and become flattened. The cells on the surface are rubbed off and are continually replaced from below. If the surface of the epithelium is dry, as on the skin, the surface cells die because the blood supply is below the basement membrane, and the scaly surfaces which develops, called *keratin*, makes a waterproof layer. If the surface is moist, as in the mouth, the cells survive until they are rubbed off, so keratin is not formed.

Transitional epithelium is like stratified epithelium but the surface cells, instead of being flattened, are rounded and can spread out when the organ expands. It is found lining organs which must expand and must be waterproof, e.g. the bladder.

Glands

Glands develop from epithelial tissues and have the ability to manufacture substances from materials brought by the blood. These substances are called the *secretions* of the gland. For example, blood contains sodium chloride and the gastric glands can make from this the hydrochloric acid found in the gastric juice, although to obtain hydrochloric acid from sodium chloride in the

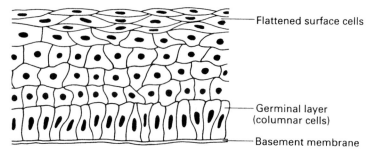

Fig. 4.2   Diagram illustrating stratified epithelium.

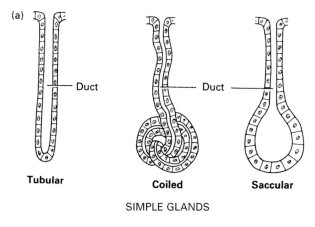

Fig. 4.3   Types of exocrine gland.

laboratory is a difficult process. Glands are well supplied with blood vessels which supply the cells with the materials necessary for making secretions. Glands are of two types: exocrine and endocrine.

Exocrine glands (Fig. 4.3) pour their external secretions out through a duct. The secretions of many of these glands contain *enzymes*, which are chemical substances produced by the cells of the gland. These enzymes produce chemical changes when they come into contact with specific substances, but do not themselves enter into the reaction.

1.   *Simple glands* have one duct leading from a single secretory unit

(a) *Simple tubular glands* are found in the walls of the small intestine, and the stomach

(b) *Simple coiled glands* pour sweat onto the surface of the skin

(c) *Simple saccular glands*, known as sebaceous glands, secrete a substance called sebum which lubricates hair and skin

2. *Compound glands* have several secretory units pouring their secretions into a number of small ducts which unite to form a larger duct

    (a) *Compound tubular glands* are found in the duodenum

    (b) *Compound saccular glands* are also called *racemose glands*, and examples are the salivary glands in the mouth

Endocrine glands pour their internal secretions directly into the blood stream. These secretions are called *hormones*. Examples of endocrine glands are the pituitary gland in the skull and the thyroid gland in the neck.

Connective Tissue

Connective tissue supports and binds together all other tissues. There are many varieties of connective tissue, which differ greatly in appearance, though there is similarity in their connective function and in the fact that all originate from primitive cells called *mesenchyme cells*, which themselves originate in the mesoderm. Connective tissue consists of cells, intercellular substance called matrix, and fibres. Matrix and fibres are non-living products of the cells, which form the supporting material of the body. The fibres are of two main types, collagenous and elastic.

Collagenous fibres originate from cells called *fibroblasts*, which secrete a substance which becomes *collagen*. These coarse fibres occur in wavy bundles and will stretch only a little without tearing.

Elastic fibres are fine branching fibres which are highly elastic.

In the body a connection sometimes needs to be firm and unyielding and sometimes a degree of elasticity is required. For example, the fibrous layers surrounding organs must be somewhat elastic to allow for swelling when the part is engorged with blood, but the fibrous tendons which join muscle to bone must be inelastic, since if they were elastic the tendon would stretch when the muscle contracted and the bone would not move. There are five main varieties of connective tissue.

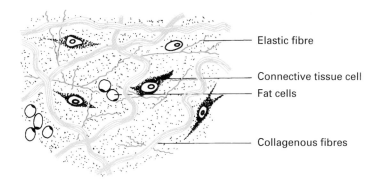

Fig. 4.4 Loose connective tissue.

Loose Connective Tissue

This type of tissue is also called *areolar tissue* and it consists of a loose network of both collagenous and elastic fibres with small scattered groups of fat cells and some fibroblasts (Fig. 4.4). Some blood vessels and nerves will be found in the tissue but these are not very numerous. Areolar tissue forms a transparent skin, as thin as tissue paper, but very tough, and it is found between and around the organs of the body.

Fatty Tissue

This is also known as *adipose tissue* and it is similar to areolar tissue but the spaces of the network are filled in with fat cells (Fig. 4.5). Fat cells contain a globule of fat, which pushes the cytoplasm and the nucleus to the edge of the cell. Adipose tissue is useful because it forms a food reserve on which the body can draw in time of need;

Fig. 4.5 Fatty or adipose tissue.

Fig. 4.6   Diagram of dense connective (fibrous) tissue.

it helps retain body heat because it is a poor conductor of heat and it protects delicate organs such as the eye and the kidney.

Dense Connective Tissue

This tissue is also called *fibrous tissue* and it consists chiefly of bundles of collagenous fibres between which the fibroblasts lie (Fig. 4.6). It is very strong compared to loose connective tissue. The fibres may be arranged regularly, in parallel bundles, as in tendons or ligaments, or irregularly, with the fibres running in different directions, as in the sheath enclosing muscles, which is called *fascia*.

Cartilage

Cartilage consists of cells, called *chondrocytes*, separated by fibres. It has no blood vessels, so the cells obtain their nourishment by diffusion of fluid through the intercellular substance. Cartilage is very tough but it is also pliant. There are three types of cartilage (Fig. 4.7):

1.   *Hyaline cartilage* consists of chondrocytes embedded in an apparently structureless matrix, which is glassy in appearance, and which has very fine collagen fibres running through it. It is found in the trachea and covering the ends of bones at a joint.
2.   *Fibro-cartilage* contains more collagen fibres than hyaline cartilage and is therefore stronger. It is found between bones forming slightly movable joints, e.g. between the bodies of the vertebrae.
3.   *Elastic cartilage* contains numerous elastic fibres embedded in

Fig. 4.7   Types of cartilage.

the matrix and can be found in the auricle of the ear and the epiglottis.

Bone

Bone is a specialized type of cartilage in which the collagen has been impregnated with mineral salts, chiefly calcium. The collagen fibres make the bone tough and the mineral salts make it rigid, so it gives proper support to the soft tissues. The cells between the fibres are called *osteocytes* and bone is richly supplied with blood vessels. The structure of bone will be dealt with more fully in Chapter 10.

Haemopoietic tissue is concerned with the formation of the blood cells and is derived from the primitive mesenchyme cells. Blood may be regarded as connective tissue with the plasma forming the matrix which contains the cells. It will be dealt with more fully in Chapter 15.

Muscular Tissue

Muscular tissue is specialized for contraction and is therefore able to produce movement. Wherever there is movement in the body there must be muscular tissue to produce it. Muscle cells are long and thin so that the shortening which occurs during contraction may be as effective as possible. Muscle cells are often called 'muscle fibres' because of their shape. The elastic fibres of connective tissue, if they have been stretched, can recoil and return to their original length, but muscle fibres can shorten themselves without this preliminary stretching. There are three types of muscular tissue: skeletal, smooth and cardiac.

Skeletal Muscle

Skeletal muscle forms the flesh of the limbs and trunk, by which the skeleton is moved. It consists of long cells often known as fibres, varying in length from a few millimetres in short muscles to from 30 cm or more in long muscles; each cell contains numerous thread-like fibres called *myofibrils*, which are only from 0·01 to 0·1 mm in width. These myofibrils are regularly striped across in alternate light and dark bands throughout their length and are so arranged that the dark and light parts come next to one another (Fig. 4.8). Each fibril is enveloped in a sheath of connective tissue called the *sarcolemma*. The fibrils are bound together by connective tissue into bundles, each enclosed in a sheath called the *endomysium*; these bundles are in turn bound together and enclosed in a

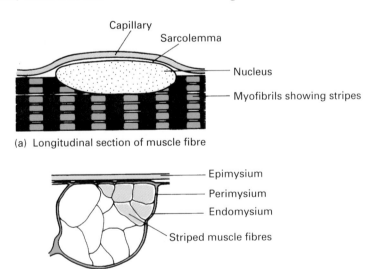

(a) Longitudinal section of muscle fibre

(b) Cross-section of part of muscle

Fig. 4.8   Sections of skeletal muscle.

sheath called the *perimysium*, to form ultimately the individual muscles, which also have a sheath of fibrous tissue called the *epimysium*. The nucleus is at the edge of the cell. This striped muscle is under voluntary control, so that it is known as 'voluntary muscle'. It contracts strongly when stimulated by a nerve fibre, but tires quickly. For strong contraction much energy is required, so voluntary muscle must have a good blood supply to bring oxygen and nutrients to the cells and to carry away waste products. Capillaries run between individual muscle cells to ensure adequate blood supply.

Smooth Muscle

Smooth muscle forms the walls of internal organs such as the stomach, bowel, bladder, uterus and blood vessels. It consists of spindle-shaped cells, each of which contains a nucleus. It is also called 'unstriped muscle'. The cells do not show any stripes and have no sheath, but are bound together by connective tissue to form the walls of the various organs (Fig. 4.9). They are not under the control of the will and act without any conscious effort or knowledge. They contract automatically, but are supplied by autonomic nerves which affect their contractions. This type of muscle is designed for slow contraction over a long period and it does not tire easily.

Cardiac Muscle

Cardiac muscle is both involuntary and irregularly striped. It is found in the heart wall only and is different from any other muscle tissue. It consists of short, cylindrical, branched fibres

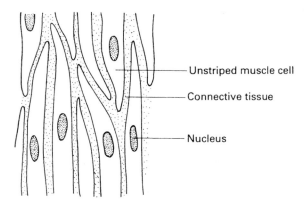

Unstriped muscle cell

Connective tissue

Nucleus

Fig. 4.9   Section of smooth muscle.

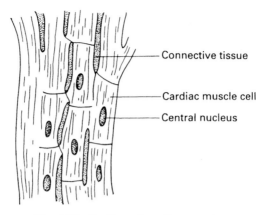

Fig. 4.10   Section of cardiac muscle.

with centrally placed nuclei. They have no sheath, but are bound together by connective tissue (Fig. 4.10). The cardiac muscle is not under the control of the will, but contracts automatically in a rhythmic manner throughout life, though the rates of these rhythmic contractions is controlled by nerves, which quicken or slow down its action. The fibres branch and join with one another so that impulses can spread from one fibre to another as well as along the length of the muscle.

Nervous Tissue

Nervous tissue is specially designed to receive stimuli from either inside or outside the body, and when stimulated to carry impulses rapidly to other tissues. Nervous tissue consists of nerve cells, called *neurones* (Fig. 4.11), and a special supporting network called *neuroglia*. A nerve cell consists of a large cell body, to which several short processes, called *dendrites*, bring impulses from other cells and tissues. From the cell body there is one long process called the *axon*, which carries impulses away from the cell body.

MEMBRANES

The cavities and hollow organs of the body are lined with membranes. These membranes consist of epithelium and secrete lubricating fluids to moisten their smooth, glistening surfaces and prevent friction. Three different types of membrane are found in the body.

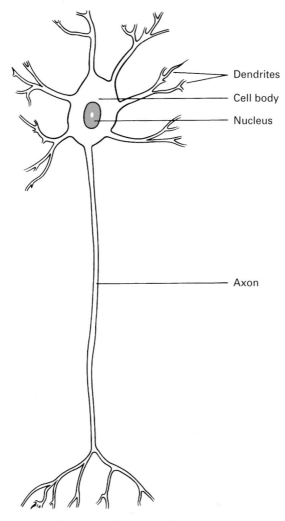

Fig. 4.11   Diagram of a neurone.

Synovial Membrane

Synovial membrane secretes a very thick fluid, which is like white of egg in consistency; hence the name, the prefix 'syn' meaning 'like' (from the Greek) and 'ovum' being Latin for an egg. It is chiefly found lining joint cavities to lubricate the movement of the bones on one another. It is a fibrous membrane covered with epithelium. It is also found over bony prominences and between ligaments and bones or tendons and bones. In these sites it forms small sacs called *bursae*, which act as water cushions, facilitating the movement of one part on another. For example, there are bursae around the shoulder, knee and elbow joints. Synovial membrane is also found forming sheaths for tendons which run a long distance to their insertions. For example, the tendons and muscles in the forearms and legs run across the hand and foot respectively and move the fingers and toes.

Mucous Membrane

Mucous membrane secretes a rather thinner fluid called *mucus*. This membrane is found lining the food canal from mouth to rectum and the air passages from the nose downwards. Thus the cavities which it lines are connected with the external skin. Mucus-secreting tubular glands are also present where much secretion is produced. These are single or branching tubes, lined with secreting cells.

Serous Membrane

Serous membrane consists of flattened cells through which oozes a small quantity of thin watery fluid called *serous fluid*; this is similar to the fluid which oozes from a blood clot. Serous membrane is found lining the internal cavities, e.g. the thorax and the abdomen, and covering the organs they contain, providing smooth, glistening, moist surfaces which slide easily over one another when one part moves on, or within, another.

ORGANS

Where more than one tissue exist together to serve a common purpose such as digesion or movement the composite structure is described as an *organ*. Examples of organs which relate to digestion and movement are the stomach and the leg, respectively. The stomach is composed of several tissue types including smooth muscle tissue, endothelial tissue, glandular tissue and nervous tissue. The leg is composed of bone, muscle, skin and nervous tissue. All organs have a blood supply.

Organs are described as being either *tubular* or *compact*. Tubular organs have certain features in common. They all contain a space called a *lumen* and the structure of the tubular structure can be broadly divided into three layers: an outer epithelial layer, a middle layer containing muscle and an inner endothelial layer. While no two tubular organs are identical in their structure these features can be seen in organs as diverse as the heart, which pumps blood round the circulatory system, and the small intestine, which transports food from the stomach to the large intestine. Each organ has unique layers, for example, in the heart the outer layer is called the *pericardium*, the middle layer is called the *myocardium* and the inner layer is called the *endocardium*. The lumen of the heart forms the *chambers* called 'atria' and 'ventricles'.

Compact organs, such as the liver and the kidney, have superficial layers called the *cortex* and deeper layers called the *medulla*. These organs have no lumen.

SYSTEMS

Where organs of the body work together towards a common purpose such as circulation, digestion or movement, this is described as a *system*. These systems will be the subject of the remainder of this book and examples of such systems are given in Box 4.2.

Systems of the body

- The **skeletal system** provides a framework which gives support and protection to the soft tissues and allows movement at joints
- The **muscular system** effects movement of the body as a whole
  The above two systems together are sometimes known as the **skeletomuscular.**
- The **circulatory system** is the transport system of the body; it carries oxygen and nourishment to the tissues and waste products away from them and is essential if there is to be dependence of one tissue on another
- The **respiratory system** allows exchange of gases between the body and the environment
- The **digestive system** is concerned with digestion and absorption of food and the elimination of waste matter
- The **endocrine system** produces hormones which control a variety of functions in the body
- The **urinary system** is the main excretory system of the body

- The **nervous system** creates awareness of the environment and makes it possible for the body to respond to change with the required precision
- The **reproductive system** is responsible for the survival of the species by reproduction of the same kind

Box 4.2

HOMEOSTASIS

The concept of *homeostasis* is fundamental to an understanding of physiology. The word homeostasis is derived from a Greek expression meaning 'staying the same'. In physiological terms it means that, despite changes in the external environment of the body, under normal healthy conditions the internal environment of the body remains the same. All the systems of the body work towards this end, and often several systems work together to maintain the constancy of a single aspect of the working of the body.

Extracellular and Intracellular Environment

One of the best examples of homeostasis is the way in which the body maintains a constant intracellular and extracellular environment with respect to water and electrolyte content. It has already been outlined (Chapter 3) how cells tend to accumulate potassium and expel sodium. Even when the body is at rest it is using energy, and most of this energy is being used by the active transport processes of cells in order to undertake this exchange of electrolytes across the plasma membrane. Other systems (discussed in later chapters) such as the *urinary system* (kidneys and bladder) are also working to expel or conserve electrolytes. The urinary system and the *respiratory system* work together to maintain the pH of the blood.

Negative Feedback

Homeostasis is enabled by means of *negative feedback*. The components of a negative feedback system are three-fold: a *detector*, an *effector* and a *control area* (Fig. 4.12). The detector senses a change in the internal environment of the body, the effector institutes a change in order to return the internal environment to homeosta-

Fig. 4.12 The concept of negative feedback.

sis, and the control area determines what the normal physiological situation should be by decreasing the action of the effector once homeostasis has been achieved.

A very good example of this (described in Chapter 17) is the regulation of blood pressure. In this case the detector is the system of pressure receptors (*baroreceptors*) in the *cardiovascular system*, the effector is a combination of actions of the heart and blood vessels and the control area is the *hypothalamus* of the brain. The link between the detector, the effector and the control area is either the nervous system or the endocrine system.

Disturbance of Homeostasis

Ill health, such as infection or injury, and many medical and surgical procedures result in disturbances of homeostasis. It can be said that nursing begins where homeostasis fails since many nursing procedures and actions are designed to detect such changes and restore the internal environment of the body.

SECTION II: CONTROL AND CO-ORDINATION

Section Introduction

Everything that the body does, including homeostasis, is controlled by the nervous and endocrine systems, and often these work together to control the internal environment. The nervous system is highly complex and a detailed consideration of the anatomical organization of the brain and spinal cord, collectively known as the central nervous system, and its links to the internal organs and musculature of the body, via the peripheral nervous system, is given.

A particular section of the peripheral nervous system which is concerned with involuntary actions, the autonomic nervous system, is also considered. Sensory information is relayed to the nervous system by different means, and two principal sets of organs, the eyes and the ears, are covered in specific chapters.

In addition to nervous signals in the body, which involve a combination of chemical and electrical signals between nerves, there is a system which is mediated wholly by means of chemical signals in the blood. The chemicals are called 'hormones' and the organs which secrete these are called 'endocrine organs'. These are, ultimately, controlled by the brain and the system whereby these are regulated and the specific action of some important hormones, such as thyroxine, are described.

Chapter 5
The Nervous System

The nervous system is:

1. The system of communication between the various parts of the body
2. The mechanism by which sensations of all kinds are received from the environment and from the tissues and organs of the body itself
3. Responsible for the interpretation of these sensations through dependence on association with similar sensations received in the past and stored in the memory
4. The system by which actions are carried out by the sending of impulses to other parts of the nervous system and other organs of the body

NERVOUS TISSUE

When nervous tissue is examined under a microscope it is seen to be composed of neurones and neuroglia (Box 5.1).

Nervous tissue

- **Neurones**—nerve cells with associated nerve fibres
- **Neuroglia**—a special type of connective tissue, found only in the nervous system, which supports the neurones

Box 5.1

The *neurone* is the basic tissue of the nervous system. It has a comparatively large *cell body*, though both shape and size vary according to the position of the cell and its function. Each has a clearly defined nucleus and the protoplasm is granular. Nerve cells form the *grey matter* of the brain and spinal cord. The cell has several processes: *dendrites* are short branched processes through which nervous impulses enter the cell and the *axon* (or axis cylinder) is a single fibre through which impulses pass out from the cell. Axons vary considerably in length from a few millimetres to many

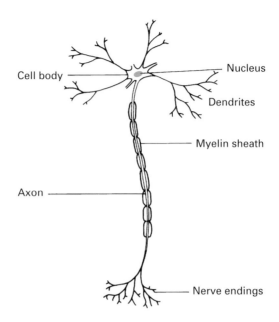

Fig. 5.1 A typical multipolar neurone.

centimetres and are continuous from cell to termination. Nerve processes, or fibres, form the *white matter* of the brain and spinal cord.

A neurone which has many processes arising from the cell body is termed a *multipolar neurone* (Fig. 5.1). Other types have one process arising from the cell body, which then divides into two branches, one conveying impulses towards the central nervous system, called the axon and one conveying impulses from an organ to the cell; these are *unipolar neurones* (Fig. 5.2). *Bipolar neurones* have two processes, one at each end of the cell; one is a dendrite carrying impulses to the cell and one is an axon conveying impulses from the cell.

Axons, and some dendrites, are surrounded by a thin fatty sheath composed of *myelin* which lies inside the outer covering of connective tissue which is called the *neurilemma* (Fig. 5.3). The myelin sheath is compressed at intervals and here the neurilemma dips in towards the nerve fibre; these constrictions are called the *nodes of Ranvier*. At these points the nerve fibre has contact with the surrounding tissue fluids, allowing exchange of nutrients and waste materials.

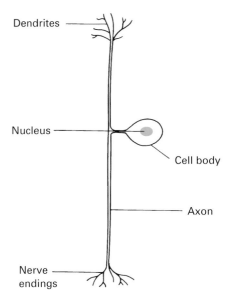

Fig. 5.2 A unipolar neurone.

The myelin sheath is thought to have an insulating effect on the nerve fibre, so that impulses are not transmitted to adjacent nerves or tissues except through the end of the fibre. It also protects the fibre from pressure and injury. Those fibres which have a myelin sheath are called *myelinated fibres*.

Non-myelinated fibres are found in the autonomic nervous system and in certain parts of the brain and spinal cord.

Nerve cells are quickly damaged by lack of oxygen, toxins and poisonous substances. If they die they cannot be replaced, though it may be possible for the function to be taken over by other cells to a limited extent.

A *synapse* is the point of communication between one neurone and the next. The fibrils forming the axon have tiny expanded ends called *end feet* which are close to, but not touching, the dendrites or cell bodies of other neurones. They allow the passage of the nerve impulse in one direction only.

Nerve impulses can also only travel in one direction, into a neurone through the cell body or dendrites and out through the axon. At the synapse there is a short pause to enable a *chemical messenger* to be released to fill the gap between the two neurones and to allow the impulse to pass to the next neurone.

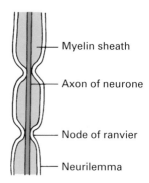

Fig. 5.3   Myelin sheath.

THE ACTION POTENTIAL

The distribution of electrolytes across all cell membranes is such that there is a *potential difference* (voltage) across the membranes. This potential difference is about −70 millivolts (mV). The cells of nerves and muscles are specialized such that they can undergo

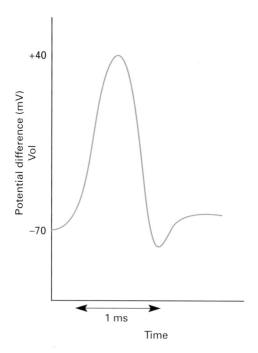

Fig. 5.4   The action potential.

reversible changes in this potential difference, called an *action potential*, and this is the basis of their ability to generate and conduct electrical impulses.

When an action potential is generated in a nerve cell, by the arrival of an electrical impulse from another nerve cell, positively charged sodium ions (Na^+) enter the cell and reverse the potential difference across the membrane to a slightly positive potential. This process is called *depolarization*. The positive potential difference is rapidly reversed (in less than a millisecond), but not before the process is repeated along the nerve cell, leading to conduction of the electrical impulse. The changes in potential difference are shown in Fig. 5.4.

Action potentials are not graded, in other words, they are not of different sizes in any particular nerve cell. They obey the *all or none law*. Once a certain level of depolarization has taken place through the entry of sodium ions into the cell, known as the *threshold potential*, a full action potential is generated. The size of action potentials does differ between nerve cells, and the extent of stimulation in a nerve (which is composed of many nerve cells) depends on the number of nerve cells which have been stimulated. A further feature of nerve cells (and muscle cells) with respect to action potentials is that they display a *refractory period* of about 0·5 milliseconds during which time another action potential cannot be generated.

The nervous system is composed of a central nervous system, consisting of the brain and spinal cord, and a peripheral nervous system, consisting of the cranial and spinal nerves and the autonomic nervous system.

CENTRAL NERVOUS SYSTEM

The Brain

The brain, when fully developed, is a large organ which fills the cranial cavity. Early in its development the brain becomes divided into three parts known as the *forebrain*, the *midbrain* and the *hindbrain* (Fig. 5.5).

The forebrain is the largest part and is called the *cerebrum*; it is divided into right and left hemispheres by a deep *longitudinal fissure* (Fig. 5.6). The separation is complete at the front and the back, but in the centre the hemispheres are joined by a broad band

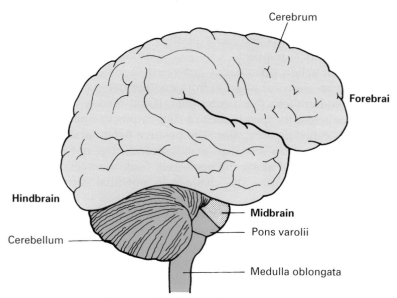

Cerebrum

Forebrai

Midbrain

Pons varolii

Medulla oblongata

Hindbrain

Cerebellum

Fig. 5.5   The brain, showing parts.

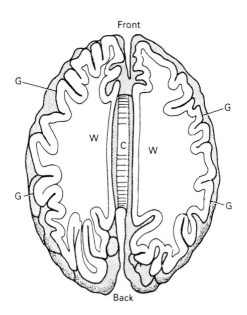

Front

G

G

W   C   W

G

G

Back

Fig. 5.6   Section through the cerebrum, viewed from above. G, The grey matter of the convoluted surface; W, the white matter forming the central portion; C, the corpus callosum, the bridge of white matter joining the two hemispheres of the cerebrum.

of nerve fibres called the *corpus callosum*. The outer layer of the cerebrum is called the *cerebral cortex* and is composed of grey matter (cell bodies) thrown into numerous folds or convolutions called *gyri*, separated by fissures called *sulci*. This enables the surface area of the brain, and therefore the number of cell bodies, to be increased greatly. The general pattern of the gyri and sulci is the same in all humans; three main sulci divide each hemisphere into four lobes, each named after the skull bone under which it lies. The *central sulcus* runs downwards and forwards from the top of the hemisphere to a point just above the lateral sulcus; the *lateral sulcus* runs backwards from the lower part of the front of the brain and the *parieto-occipital sulcus* runs downwards and forwards for a short way from the upper posterior part of the hemisphere. The lobes of the hemispheres are: the *frontal lobe*, lying in front of the central sulcus and above the lateral sulcus; the *parietal lobe*, lying between the central sulcus and the parieto-occipital sulcus and above the line of the lateral sulcus; the *occipital lobe*, which forms the back of the hemisphere; and the *temporal lobe* lying below the lateral sulcus and extending back to the occipital lobe (Fig. 5.7).

The area lying immediately in front of the central sulcus is known as the *pre-central gyrus* and is the motor area from which arise many of the motor fibres of the central nervous system. Immediately behind the central sulcus lies the sensory area, called the *post-central gyrus*, in the cells of which several kinds of sensation are interpreted.

Longitudinal section of a hemisphere shows grey matter (cell

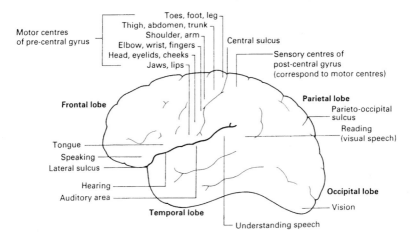

Fig. 5.7   The cerebrum, showing the lobes and the main nerve centres.

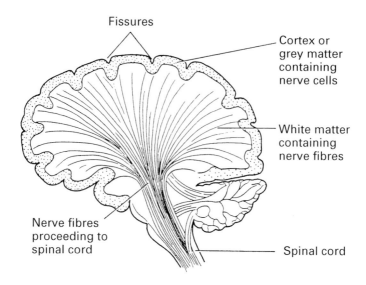

Fissures

Cortex or
grey matter
containing
nerve cells

White matter
containing
nerve fibres

Nerve fibres
proceeding to
spinal cord

Spinal cord

Fig. 5.8   Diagrammatic section of the brain, showing the grey matter on the surface and the white matter in the centre. The association fibres in the cerebrum are not shown.

bodies) on the outside and white matter (nerve fibres) forming the interior (Fig. 5.8). The nerve fibres connect one part of the brain with other parts and with the spinal cord, but within the white matter groups of nerve cells can be seen forming areas of grey matter. These areas of grey matter are called *cerebral nuclei* (Fig. 5.9). The main function of these areas is co-ordination of movement and posture of the body: disorders affecting these areas cause jerky movements and unsteadiness.

The cavities within the brain are called *ventricles*. There are two *lateral ventricles*, a central *third ventricle* and a *fourth ventricle* between the cerebellum and the pons. All are filled with cerebro-spinal fluid.

The midbrain lies between the forebrain and the hindbrain. It is about 2 cm in length and consists of two stalk-like bands of white matter called the *cerebral peduncles*, which convey impulses passing to and from the brain and spinal cord, and four small prominences called the *quadrigeminal bodies*, which are concerned with sight and hearing reflexes. The *pineal body* lies between the two upper quadrigeminal bodies.

The hindbrain has three parts:

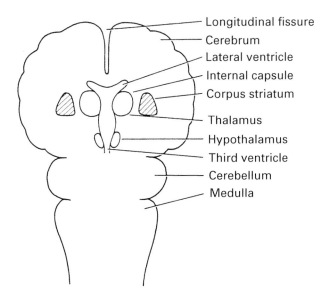

Longitudinal fissure
Cerebrum
Lateral ventricle
Internal capsule
Corpus striatum
Thalamus
Hypothalamus
Third ventricle
Cerebellum
Medulla

Fig. 5.9 Cerebral nuclei.

1. The *pons*, which lies between the midbrain above and the medulla oblongata below. It contains fibres which carry impulses upwards and downwards and some which communicate with the cerebellum.

2. The *medulla oblongata* lies between the pons above and the spinal cord below. It contains the cardiac and respiratory centres which are also known as the vital centres and which control the heart and respiration.

3. The *cerebellum* projects backwards beneath the occipital lobes of the cerebrum. It is connected to the midbrain, the pons and the medulla oblongata by three bands of fibres called the *superior, middle* and *inferior cerebellar peduncles* respectively. The cerebellum is responsible for the coordination of muscular activity, control of muscle tone and maintenance of posture. It is continuously receiving sensory impulses concerning the degree of stretch in muscles, the position of joints and information from the cerebral cortex. It sends information to the thalamus and the cerebral cortex.

The midbrain, the pons and the medulla have many functions in common and together are often known as the *brain stem*. This area also contains the nuclei from which originate the cranial nerves.

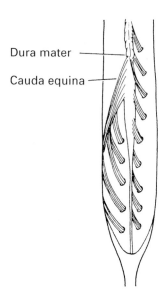

Dura mater

Cauda equina

Fig. 5.10   The cauda equina.

Spinal Cord

The spinal cord is continuous with the medulla oblongata above and constitutes the central nervous system below the brain. It commences at the foramen magnum and terminates at the level of the first lumbar vertebra; it is about 45 cm in length. At its lower end it tapers off into a conical shape called the *conus medullaris* from the end of which the *filum terminale* descends to the coccyx, surrounded by nerve roots called the *cauda equina* (Fig. 5.10). The

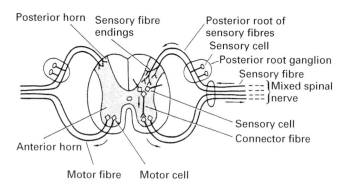

Fig. 5.11   A section of the spinal cord, showing the central grey matter and the neurones. Sensory fibres of heat and pressure cross the cord.

cord gives off nerves in pairs throughout its length. It varies somewhat in thickness, swelling out in both the cervical and lumbar
regions, where it gives off the large nerve supply to the limbs. The
cord is deeply cleft back and front, so that it is almost completely
divided into right and left sides like the cerebrum.

The cord, like the medulla, consists of white matter on the
surface and grey matter in the centre. The white matter consists
of fibres running between the cord and the brain only, not to the
body tissues (Fig. 5.11). The cord contains *motor fibres* and *sensory
fibres* (Box 5.2).

Composition of white matter

- **Motor fibres** run down from the motor centres of the cerebrum and the cerebellum to the motor cells of the cord
- **Sensory fibres** run up the cord from the sensory cells of the
  cord to the sensory centres of the brain

Box 5.2

The *grey matter*, on cutting across the cord, has an H-shaped
pattern, with two portions projecting forwards, one on either
side, called the *anterior horns*, and two portions projecting
backwards, one on either side, called the *posterior horns*.

The *cranial nerves* are twelve pairs of nerves which arise in nuclei
in the brain stem. Some are purely sensory, some purely motor and
some are mixed nerves carrying both sensory and motor impulses.

The *spinal nerves* are 31 pairs of nerves which arise in the spinal
cord. Each has a motor and a sensory component in the anterior
and posterior parts of the cord respectively, and the two fibres
travel together after they leave the cord.

The *autonomic nervous system* is concerned with the control of
internal organs; the function of these organs is not under the
control of the will, though they are affected by the emotions.

The cranial and spinal nerves and the autonomic nervous system, which as mentioned earlier form the peripheral nervous
system, are discussed in greater detail later in this chapter.

The Motor System

The motor system is concerned with movement of various parts of
the body. As already mentioned (p. 63), the *motor area* is situated in

front of the central sulcus in an area called the *pre-central gyrus*. Here the body is represented upside down. At the bottom of the gyrus is a large area for the head and eye; above that is a large area for the hand and arm, then a small area for the trunk and a large area for the leg extending over the top of the cerebral hemisphere (see Fig. 5.7). There is considerable overlap between these areas. It will be apparent that an area which can undertake a great deal of fine movement such as the hand and arm, will require a larger area of cell bodies than an area such as the trunk which, although greater in extent, does not carry out as much detailed movement. In front of the motor area lies the *pre-motor area* which is concerned with a whole pattern of movement.

Beginning from cells in the motor area the corticospinal fibres pass downwards in a fan-shape (see Fig. 5.8) and then pass through the *internal capsule*, which lies between the thalamus and the basal ganglia. All the motor fibres serving one side of the body are gathered together in the internal capsule so injury there will cause paralysis in the affected side (hemiplegia). The fibres then pass through the pons to the medulla oblongata where they form a long narrow projection called a *pyramid*. In the pyramids most of the motor fibres cross over to the other side, at the *decussation of the pyramids*, so that fibres that arose in the left cerebral hemisphere will now be on the right side of the tract and will supply the right side of the body. The fibres then pass down the spinal cord as the *lateral corticospinal tract*. The fibres which did not cross to the opposite side at the decussation of the pyramids pass down the spinal cord in the *anterior corticospinal tract* and eventually cross to the opposite side in the spinal cord.

Fig. 5.12   Muscle, showing motor end plates and sensory nerve endings.

The fibres pass to the anterior horn of the spinal cord where they form a synapse with the cell bodies situated there. The fibres then emerge from the front of the spinal cord as the *anterior root* and join the corresponding posterior root of sensory fibres to form a *mixed spinal nerve* (see Fig. 5.11). As peripheral nerves these end in branches to various areas, including muscles. The motor fibres divide into branches and each branch ends in a *motor end plate* attached to an individual muscle fibre (Fig. 5.12). Sensory fibres have their cells in the posterior root ganglion on the spinal nerves and have endings of various types within the muscles.

The motor area receives impulses from many other parts of the brain, including the sensory area. From the cortex impulses are sent to the spinal cord, the motor nuclei in the brain stem, the basal ganglia, the cerebellum and the pons. Through the various nerve tracts stimuli are passed through the peripheral nerves to the skeletal muscles, which are kept in a state of tension called *muscle tone*. Muscle fatigue is prevented by the use of different sets of motor fibres successively and the degree of tone depends on the number of fibres being used at any one time.

The term *upper motor neurone* describes the motor fibre within the central nervous system as far as its synapse with an anterior horn cell. A *lower motor neurone* describes an anterior horn cell and its fibre.

The corticospinal tract used to be called the pyramidal tract, so the term *extra-pyramidal system* describes all motor systems other than the corticospinal and corticonuclear tracts. The main function of the system is co-ordination of muscle movement in the maintenance of body posture so that movements can be performed accurately while still maintaining the desired posture.

The Sensory System

The sensory system is concerned with interpreting the impulses which are constantly stimulating it. Many of these do not reach the level of consciousness because the nervous system deals with them automatically making adjustments to the height of the blood pressure, the rate of the heart beat, the degree of tone in the muscles and many other conditions.

Sensory impulses are transmitted to the central nervous system from the special sense organs, the skin, and from deeper parts of the body.

Special Senses

The impulses of *sight* are conveyed from the retina of the eye through the *optic nerve* (second cranial nerve) to the *optic chiasma* where the medial fibres of each optic nerve cross over to the opposite side (Fig. 5.13). Because of this partial crossing the visual area of the left cerebral hemisphere receives impressions from the outer (temporal) side of the retina of the left eye and from the inner (nasal) side of the retina of the right eye. The impulses are then conveyed to the *visual areas* in the occipital lobes, where they are interpreted. It will be seen that division of the left optic nerve causes blindness of the eye on the same side but that division of the left optic tract causes inability to see the left half of the normal field of vision.

The impulses of *hearing* are conveyed through the vestibulo-cochlear nerve (eighth cranial nerve) to the *pons* where there is a synapse. A second set of fibres carry the impulses to the *medial geniculate body* and a third set to the *auditory areas* in the temporal lobes for interpretation.

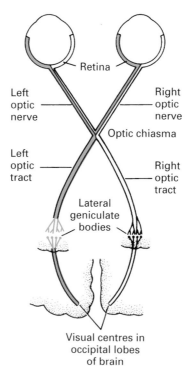

Fig. 5.13   The visual path from the eyes to the visual centres.

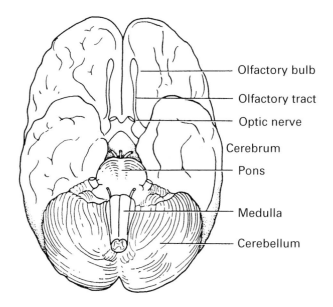

Olfactory bulb

Olfactory tract

Optic nerve

Cerebrum

Pons

Medulla

Cerebellum

Fig. 5.14 The brain, viewed from below.

The impulses of *smell* are carried to the *olfactory nerve* (first cranial nerve) and then to the *olfactory bulb* and *olfactory tract* which are under the frontal lobe (Fig. 5.14). Smells are interpreted in various parts of the temporal lobe. There are few *tastes* which can be detected without the sense of smell. Taste impulses are carried by the facial (seventh cranial) and glossopharyngeal (ninth cranial) nerves and are interpreted in the temporal lobe with the corresponding smell.

Sensation from Skin and Muscle
Sensory nerve endings are present in the skin and other tissues. Different kinds of sensation—temperature, touch, pressure and degrees of stretch—require different nerve endings to initiate them. In the skin a sensory nerve fibre may begin as: (1) a nerve ending capable of transmitting pain and temperature change, (2) one transmitting light touch, or (3) one transmitting firm pressure. In addition muscles have special structures called *muscle spindles* which respond to the degree of tension to which they are exposed. From all these nerve endings sensory fibres run in the spinal nerves to the posterior nerve roots in the spinal cord. The nerve cell of the first neurone is in the posterior root ganglion. In the head they run

in the trigeminal nerve (fifth cranial) and other cranial nerves to the brain.

Nerve fibres carrying light touch and muscle stretch sensations (proprioception) then give off several branches to anterior and posterior horn cells so that each section of the cord becomes a functioning unit. They then pass upwards in the posterior part of the white matter and end in a synapse in the medulla oblongata. A second neurone crosses over to the opposite side and ends at the thalamus. A third neurone carries impulses to the sensory area in the parietal lobe.

Fibres carrying pain and temperature change synapse in the posterior horn. A second neurone crosses over immediately to the opposite side of the cord and ascends to form a synapse in the thalamus. A third neurone passes to the sensory area.

Fibres carrying firm pressure impulses also synapse in the posterior horn. A second neurone crosses over to the opposite side of the cord and ascends, in a different part of the cord, to the thalamus. A third neurone passes to the sensory area.

It can be seen that all sensory fibres eventually cross to the opposite side so that sensation from the left side of the body will be interpreted in the right side of the brain. Also, all sensory neurones form a synapse in the thalamus.

The *thalamus* (see Fig. 5.9) is responsible for sorting out the mass of sensory information being fed into the body and passing it on to the cerebral cortex, when necessary, or to other areas of the brain as appropriate. The *hypothalamus* is concerned with the stability of the internal organs of the body. It controls water balance, regulates appetite, temperature and sleep and plays a part in the control of emotion.

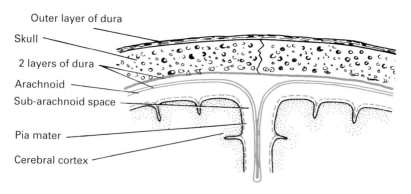

Fig. 5.15   Diagram of the meninges.

The sensory area of the cortex lies in the parietal lobe behind the central sulcus. Like the motor area the body is represented upside down, with large areas for the face, head and hand at the lower end and the smaller areas for the arm, trunk and leg which are comparatively less sensitive.

The Meninges

The meninges are the protective membranes covering the central nervous system (Fig. 5.15). There are three layers (Box 5.3).

Meningial layers

- **Dura mater**—outer layer
- **Arachnoid mater**—middle layer
- **Pia mater**—inner layer

Box 5.3

The outer layer is called the *dura mater*. It is a tough fibrous membrane which has two layers, the outer layer lining the inner surface of the skull and forming the periosteum. At the foramen magnum this layer continues as the periosteum on the outer surface of the skull. The inner layer of dura projects inwards in certain places to form a double layer which separates parts of the brain and helps to maintain them in position. The *falx cerebri* is one such fold, between the two cerebral hemispheres; another is the *tentorium cerebelli* between the cerebrum and the cerebellum. The two layers of the dura are in contact with one another for the most part, but are separate when they enclose a venous sinus. The inner layer of dura mater also encloses the spinal cord and continues down as far as the sacrum.

The *sub-dural space* is a potential rather than an actual space which lies between the dura mater and the arachnoid mater.

The *arachnoid mater* is a delicate membrane which lies immediately beneath the dura and dips down with it between the main portions of the brain.

The *sub-arachnoid space* lies between the arachnoid mater and the pia mater and is filled with cerebrospinal fluid. Between the cerebellum and the medulla oblongata there is a comparatively large space called the *cisterna magna*. This is used to obtain a sample of cerebrospinal fluid in small children. The arachnoid mater accompanies the dura as a covering to the spinal cord and extends to the sacrum.

The *pia mater* is a thin, vascular membrane which is in contact with the surface of the brain and spinal cord and dips into all the convolutions.

The Cerebrospinal Fluid

The cerebrospinal fluid is a clear, colourless liquid which fills the sub-arachnoid space and the ventricles of the brain. It is secreted by the *choroid plexuses* in the ventricles, and passes from the two lateral ventricles, which communicate with each other and with the third ventricle through the interventricular foramen, to the third ventricle and then through a narrow tube, called the *aqueduct*, into the fourth ventricle. There are three small openings in the roof of the fourth ventricle through which the cerebrospinal fluid passes into the sub-arachnoid space in which it circulates around the outside of the brain and spinal cord. It is finally absorbed through the *arachnoid granulations*, which are small projections of arachnoid mater, into the venous sinuses.

The cerebrospinal fluid is similar to blood plasma in composition although it has only a small amount of protein. There are about 120 ml of fluid altogether with a pressure of 60 to 150 mm of water. It usually contains 200–300 mg protein per litre and about 2·8–4·4 mmol glucose per litre. These amounts may be altered in disease.

The main function of the cerebrospinal fluid is to protect the brain and spinal cord by forming a water cushion between the delicate nerve tissues and the walls of the bony cavities in which they lie. It enables the pressure within the skull to be kept constant and it carries away waste and toxic substances.

PERIPHERAL NERVOUS SYSTEM

The Cranial Nerves

The cranial nerves originate or terminate within the brain. Some are motor nerves, some are sensory nerves, and some are mixed nerves (see Table 5.1).

Table 5.1 Cranial Nerves

Name	Type	Function and distribution
1. Olfactory nerve	Sensory	The nerve of smell. Starts in the nose and passes to the olfactory bulb
2. Optic nerve	Sensory	The nerve of sight. Starts in the retina and passes to the lateral geniculate body
3. Oculomotor nerve	Motor	Arises in the midbrain and ends in the muscles which move the eye
4. Trochlear nerve	Motor	As the third cranial nerve
5. Trigeminal nerve	Motor and sensory	Supplies the muscles of mastication and has three sensory branches—ophthalmic, maxillary and mandibular
6. Abducent nerve	Motor	Arises in the pons and ends in one of the muscles moving the eye
7. Facial nerve	Motor and sensory	Supplies the muscles of facial expression and is sensory from the tongue
8. Vestibulocochlear nerve	Sensory	Branches from the ear and the cochlea give hearing and the sense of balance
9. Glossopharyngeal nerve	Motor and sensory	The nerve of taste. Also sends motor fibres to the pharynx
10. Vagus nerve	Motor and sensory	Supplies the digestive tract controlling both secretion and movement
11. Accessory nerve	Motor	Supplies the muscles of the neck, most of the pharynx and soft palate
12. Hypoglossal	Motor	Supplies the tongue

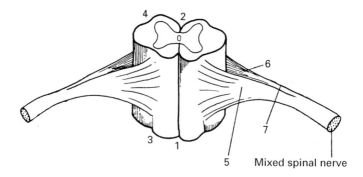

Fig. 5.16 A section of the spinal cord, showing a pair of spinal nerves arising from it. (1, 2) Fissures which divide the right and left sides of the cord. (3, 4) Smaller fissures at the side from which the nerves arise. (5) The anterior root of the spinal nerve. (6) The posterior or sensory root, on which the posterior root ganglion (7) shows as a swelling close to the point where the two roots join.

The Spinal Nerves

The spinal nerves are divided into groups according to the region of the cord from which they arise (Fig. 5.16; Box 5.4).

Groups of spinal nerves

- Eight pairs of **cervical** nerves, one above the atlas and one below each of the cervical vertebrae
- Twelve pairs of **thoracic** nerves, one below each thoracic vertebra
- Five pairs of **lumbar** nerves ⎫
- Five pairs of **sacral** nerves ⎬ derived from the cauda
- One pair of **coccygeal** nerves ⎭ equina

Box 5.4

The spinal nerves give off short posterior branches which supply the muscles of the back of the neck and trunk, and long anterior branches, which provide the nerves of the limbs and the sides and front of the trunk.

In certain regions these nerves branch immediately they leave the spinal canal, and the branches join up with one another to form the nerves supplying the various muscles and parts. This inter-branching is called a *plexus*. Plexuses are formed in all regions except the thoracic region.

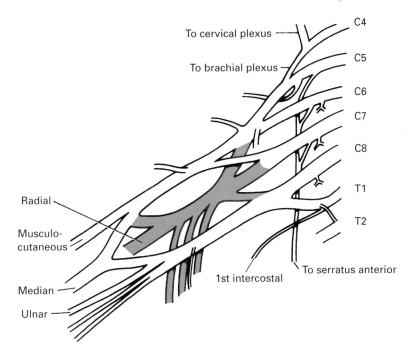

To cervical plexus

To brachial plexus

C4

C5

C6

C7

C8

T1

T2

Radial

Musculo-
cutaneous

Median

Ulnar

1st intercostal

To serratus anterior

Fig. 5.17 The brachial plexus and the main nerves arising from it.

The *cervical nerves* form two plexuses:

1. The *cervical plexus*, which supplies muscles of the neck and shoulder, and also gives off the *phrenic nerve* supplying the diaphragm.
2. The *brachial plexus*, which supplies the upper limb.

The *brachial plexus* gives off three main nerves—the radial, median and ulnar nerves (Fig. 5.17). The *radial nerve* runs round the back of the humerus and down the outer side of the forearm. It supplies the extensor muscles of the elbow, wrist and hand. It is exposed to pressure in the armpit and against the humerus, and injury to it is the cause of wrist-drop, when the joint is flexed and cannot be extended. This may result from the use of badly padded, cheap crutches, with no hand rest, pressing in the armpit or from pressure by the edge of the operating table on the nerve against the humerus, if the patient's arms are allowed to hang down during an operation.

The *ulnar* and *median nerves* run down the inner side and middle of the limb respectively, and supply the flexor muscles of the wrist

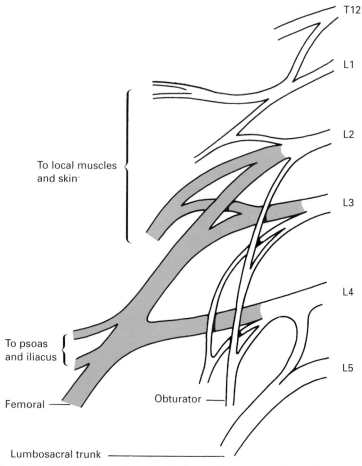

Fig. 5.18   The lumbar plexus and the main nerves arising from it. The Lumbosacral trunk joins the sacral plexus.

and hand. Injury to them causes hyperextension and gives rise to the 'claw-like' hand, the unopposed extensor muscles coming into play. A fourth smaller nerve, the *musculocutaneous nerve*, supplies the flexors of the elbow-joint, the biceps, and brachialis muscles. It is the ulnar nerve, crossing in the groove between the back surface of the internal epicondyle of the humerus and the olecranon, which is knocked when we say we have 'knocked our funny-bone', the tingling pain passing down the nerve to the hand.

The *thoracic nerves* supply the muscles of the chest and the main part of the abdominal wall.

The *lumbar nerves* form the lumbar plexus, which gives off one main nerve, the *femoral nerve* (Fig. 5.18). This runs down beside the psoas muscle under the inguinal ligament into the front of the thigh and supplies the muscles there. The lumbar plexus also gives off branches to the lower plexus wall.

The *sacral nerves* with branches from the fourth and fifth lumbar nerves form the sacral plexus, which gives off one large nerve, the *sciatic nerve* (Fig. 5.19). This is the largest nerve in the body. It leaves the pelvis by the sciatic notch, runs across the back of the hip-joint and down the back of the thigh, supplying the muscles there. It divides above the knee into two main branches:

1.   The *peroneal nerve*, which supplies the muscles of the front of the leg and foot.

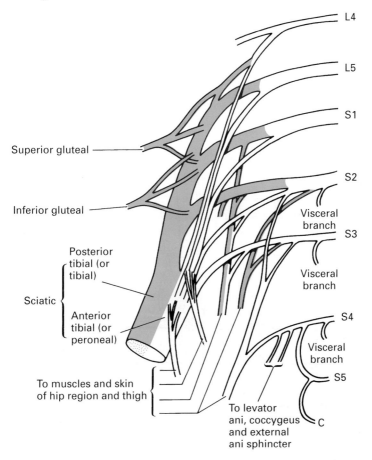

Fig. 5.19   The sacral plexus, showing the main nerves arising from it.

Fig. 5.20 Diagrammatic representation of the parts of the autonomic nervous system. The sympathetic nervous system is shown by the continuous lines, and the parasympathetic nervous system by broken lines.

2.   The *tibial nerve* which supplies the muscles of the back of the leg.

The sciatic nerve therefore supplies the whole of the leg below the knee except for a small sensory branch from the femoral nerve.

The *coccygeal nerves*, with branches from the lower sacral nerves, form a second small plexus on the back of the pelvic cavity, supplying the muscles and skin in that area, e.g. the muscles of the perineal body, the external sphincter of the anus, the skin, and other tissues of the external genitals and the perineum, etc.

Both the sacral and the coccygeal nerves also give off branches to the sympathetic ganglia in the pelvic area.

The Autonomic Nervous System

The autonomic nervous system supplies nerves to all the internal organs of the body and the blood vessels (Fig. 5.20). It is so named because these organs are self-controlled (auto = self) and not under the control of the will. The functioning of the internal organs normally takes place without any conscious knowledge. The will does not normally affect them, but the emotions do. They are affected by the hypothalamus.

The autonomic nervous system is for the most part efferent. It is made up largely of efferent neurones, both motor, supplying the involuntary muscles of the walls of organs such as the stomach, intestines, bladder, heart and blood vessels, and secretory, supplying the glands such as the liver, pancreas and the kidney. There are some afferent fibres, but they are comparatively few in number, as the internal organs are almost insensitive. As a result disease may attack and destroy them without causing pain, and the pain which does occur is due to a large extent to inflammation of the lining membranes of the cavity in which they lie; for example, tuberculosis or pneumonia may affect the lung tissue without any pain, but as soon as the pleura is involved sharp pain is felt. In the same way it hurts to cut through the abdominal wall, but when a piece of bowel has been brought out of the abdomen it can be cut without causing the patient any pain, as the nurse may see in cases of colostomy.

The autonomic nervous system consists of two parts (Box 5.5).

The *sympathetic nervous* system consists of a *double chain of ganglia* running down the trunk just in front of the vertebral column in the cervical, thoracic and lumbar regions. These ganglia are linked to one another by nerves. They receive nerves from the thoracic and

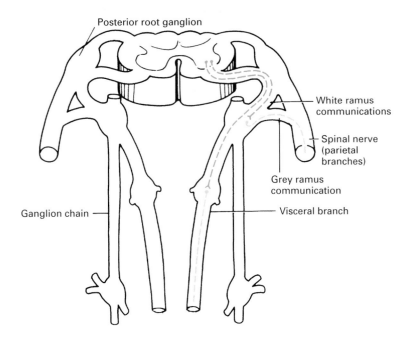

Fig. 5.21 Section of the spinal cord, showing the efferent and afferent pathways of the sympathetic nerves and the link with the spinal nerve. (After Cockett.)

Divisions of the autonomic nervous system

- The sympathetic nervous system
- The parasympathetic nervous system

Box 5.5

---

upper lumbar regions of the spinal cord (Fig. 5.21). They give off nerves supplying the internal organs, the *visceral branches*, and nerves running back to the spinal nerves, the *parietal branches*. These nerves supply the blood vessels, the sweat and sebaceous glands and the muscles which raise the hairs of the skin making them 'stand on end'. In certain regions where there are many organs requiring a nerve supply there are additional ganglia between the two chains, linked up by nerves with the chains and with one another, giving off nerves to the neighbouring organs; these are called plexuses, e.g. the cardiac plexus lies behind the heart in the thoracic region, and the solar plexus lies just below the diaphragm, where stomach, liver, kidneys, spleen and pancreas are all found.

The *parasympathetic* nervous system consists chiefly of the vagus nerve, which gives off branches to all the organs of the thorax and abdomen, but does also include branches from other cranial nerves (the third, seventh and ninth), and nerves from ganglia in the sacral region of the vertebral column.

The Functions of the Autonomic System

All the internal organs have, therefore, a double nerve supply, sympathetic and parasympathetic, and the two sets of nerves have in each case opposite actions, one *stimulating* and the other *checking* the activity of the organ. The arrangement is similar to a motorcar, with an accelerator pedal to make it move faster and a brake pedal to check it.

The *sympathetic* nerves have a stimulating and quickening effect on the heart and on the respiratory system, but a checking effect on digestion. They improve the circulation and cause dilation of the bronchial tubes, increasing the air intake, but they stop the secretion of digestive juices from the salivary glands, and throughout the alimentary canal, and check peristaltic action in its wall. These nerves are stimulated by strong emotion, such as fear, anger and excitement. (It is because of this effect of the emotions that they are called sympathetic.) Their functions are thus closely linked to those of the adrenal medulla, which they stimulate. They help to enable the body to respond to emotion, since they provide the muscles with a better supply of blood which is rich in oxygen. This enables the individual to run away when frightened or to fight when angry, the instinctive response to these emotions. On the other hand, they are responsible for the arrest of the digestion of food in strong emotion, and thus may produce vomiting and emptying of the bowel, the organs getting rid of the contents which they cannot cope with for the time being.

The *parasympathetic* nerves have exactly opposite effects, stimulating the digestive system and producing both a copious flow of digestive juices and peristaltic action. On the other hand, the vagus slows the heart, reducing the circulation, and has a checking effect on the respiratory system, contracting the bronchial tubes. These nerves are stimulated by pleasant emotions. As a result, happiness and a contented mind tend to improve digestion. Pavlov was able to show this in a dog with a gastric fistula. When the dog was shown a bone which pleased him, gastric juice began to pour into the stomach by reflex action through the vagus nerve. The bringing of a cat into the room angered the dog and the flow was checked, the sympathetic nerves being stimulated. Pavlov also noticed that if a bell was rung before the dog's food was brought in each day, after a time the ringing of the bell without bringing the food would cause a flow of gastric juice. This is termed a *conditioned reflex*. The animal had learned to associate the two things, so that either would produce the same reflex response. It is the same with human beings. The ringing of a dinner bell will cause secretion of digestive juices, just as will the sensation of the smell and sight of pleasant, well-served food. It is therefore of special importance to serve foods appetizingly, and choose dishes which will give the patient as much pleasure as possible.

REFLEX ACTIONS

Reflex action is the result of the stimulation of the motor cells by stimuli brought in by afferent neurones from the tissues. Incoming stimuli can therefore, in addition to causing sensation, give rise to action. They only produce sensation if they are passed on to the sensory centres of the brain. On the other hand, in cord and brain they may stimulate the motor cells and give rise to action—*reflex action* (Fig. 5.22). Sensory stimuli are pouring into the cord and brain from the tissues all the time. If they reach the sensory centres of the cortex of the cerebrum and stimulate them they produce sensations of which we are conscious. If they stimulate motor cells they produce reflex action, e.g. the touch of something hot on the skin causes immediate withdrawal of the part, a tap on the patella ligament causes contraction of the quadriceps extensor muscle and produces the 'knee-jerk', stroking of the sole of the foot causes the toes to be drawn down. In the case of the first action it may be claimed that because the heat causes an unpleasant sensation we want to withdraw the part and the action is voluntary. On the other hand, the action occurs in an animal when the spinal cord has been severed in the neck, so that no stimuli can reach the brain, and is, in fact, more marked than in an uninjured animal. The sensory stimulus is brought into the cord by sensory fibres, transmitted by connector fibres to the motor cells of the anterior horn, and passed out by the motor fibres to the muscles. In the case of the knee-jerk the sensory stimulus is carried into the lumbar region of the cord by afferent fibres and stimulates the motor cells, which control the quadriceps.

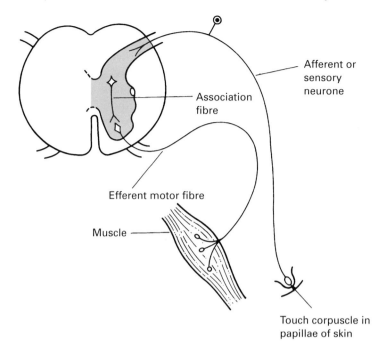

Fig. 5.22 A reflex arc.

The reason for the reflex being more marked when the cord is cut off from the brain is that the cerebral centres have an *inhibiting effect* on reflex action. Reflex action is not produced by the will; it is a mere response to environment, and is the only type of action found in the lower forms of life. With the development of the cerebrum, control of reflex action and the partial replacement of reflex action by voluntary action takes place. If we touch something hot the natural reflex is to withdraw, as the inexperienced puppy does when it touches a hot cinder in the grate. On the other hand, if we pick up a hot plate with our dinner on it, appreciating the value of the plate and the dinner, we can continue to hold the plate and put it safely down at the expense of burnt fingers. The reflex is inhibited. If we have been warned and expect the plate to be hot, this control is more easily established. In the case of the pure reflex, where the action which is produced has no value, as in the knee-jerk and the drawing down of the toes on tickling the sole of the foot, there is still some inhibiting effect from the brain, and if the nerve path from the brain to muscle is destroyed by injury or disease above the anterior horn cell the action is much more marked.

Actions which are in the first place voluntary become in a sense reflex, e.g. standing is in the first place a voluntary act which is carried out by the exercise of the will. When we have learnt to keep our balance on two feet we learn to do it by the sensations from the skin of our feet, and from muscles and joints, and the sensory organs of balance, and we can stand without voluntary effort unless disease affects the sensory nerves. In the same way when we learn to knit it is at first a voluntary action. Gradually our fingers and arms learn the feel of the wool, needles and actions, and we can knit without attention unless we lose the sensation in the parts or are learning a new complicated pattern.

Reflex action may occur at three different levels in the nervous system:

1. The spinal reflex, e.g. the knee-jerk
2. Reflexes occurring at the base of the brain, e.g. sneezing, coughing, vomiting, walking (cerebellar)
3. Reflexes occurring in the cerebrum, and involving use of the association fibres of the brain

SUMMARY

In this chapter the major structural features of the nervous system have been covered. The nervous system is divided into the central and peripheral systems, and the subdivisions of the peripheral system are the cranial and spinal nerves and the autonomic nervous system. Some key features of function, such as the generation and transmission of nerve impulses and the reflex arc, have also been covered.

Chapter 6
The Ear

The ear is the organ of hearing and also plays an important part in the maintenance of balance. The external ear, the middle ear and the cochlea of the internal ear are concerned with hearing; the semicircular canals, the utricle and the saccule of the internal ear are concerned with balance.

THE STRUCTURE OF THE EAR

The *external ear* has two parts: the auricle and the external acoustic meatus (Fig. 6.1). The *auricle* projects from the side of the head. It is composed of a thin piece of elastic fibro-cartilage, covered with skin, which funnels sound waves towards the external acoustic meatus. The *external accoustic meatus* is a tubular passage about

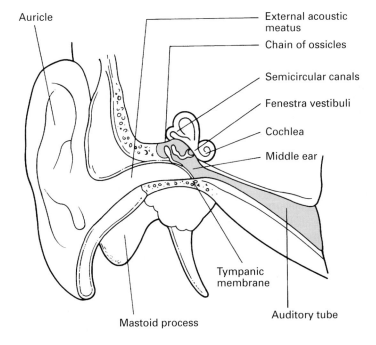

Fig. 6.1 The ear.

4 cm long leading into the temporal bone. The outer one-third has walls of cartilage and the inner two-thirds walls of bone; the canal is curved, running first forwards and upwards, then backwards and upwards and finally forwards and slightly downwards. These curves may be straightened by gentle traction on the auricle; in adults the pull is upwards and backwards, in children backwards only and in infants downwards and backwards. The inner end of the meatus is closed by the *tympanic membrane*. The skin lining the cartilaginous meatus contains hair follicles and numerous glands which secrete *cerumen*. These protect the canal from foreign bodies by entangling dust and other particles, but the cerumen may itself block the canal if it accumulates, and will then require removal by syringing.

The *middle ear* is a small space within the temporal bone. The tympanic membrane separates it from the external ear and its further (medial) wall is formed by the lateral wall of the internal ear. The cavity is lined with mucous membrane and is filled with air which enters from the pharynx through the auditory tube. This equalizes air pressure on both sides of the tympanic membrane. It contains a chain of three tiny bones, called *ossicles*, which transmit the vibrations of the tympanic membrane across to the internal ear. The tympanic membrane is thin and semitransparent and the handle of the *malleus*, the first of the ossicles, is firmly attached to the inner surface. The *incus* articulates with the malleus and with the *stapes*, the base of which is attached to the fenestra vestibuli, which leads to the internal ear. The posterior wall of the middle ear has an irregular opening which leads into the *mastoid antrum* and this in turn leads to a number of mastoid air cells. These are air-filled cavities within the bone which, like the nasal sinuses, may become infected.

The *internal ear* lies in the petrous part of the temporal bone. It consists of two parts: the bony labyrinth and the membranous labyrinth.

The *bony labyrinth* is again divided into three parts: the vestibule, the cochlea and the semicircular canals (Fig. 6.2).

The *vestibule* adjoins the middle ear through two openings; the *fenestra vestibuli*, which is filled by the base of the stapes, and the *fenestra cochlea* which is filled by fibrous tissue. At the back are openings into the semicircular canals and at the front an opening into the cochlea.

The *cochlea* (Fig. 6.3) is concerned with hearing. It is a spiral tube which makes two and three-quarter turns round a central pillar of bone called the *modiolus.* The tube is divided lengthwise into three

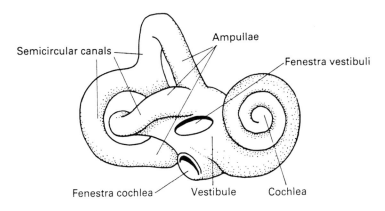

Fig. 6.2 The bony labyrinth.

separate tunnels by two membranes: the basilar membrane and the vestibular membrane, which stretch from the modiolus to the outer wall. The outer tunnels are the scala vestibuli above and the scala tympani below. These tunnels are filled with perilymph and they join at the top of the modiolus. The lower end of the scala tympani is closed by the fibrous fenestra cochlea. The middle tunnel is called the cochlear duct and is filled with endolymph. It is the same shape as the bony labyrinth and is called the membranous labyrinth. Within the cochlear duct are the special nerve endings of the auditory nerve called hair cells.

The *semicircular canals* are three in number and they lie above and behind the vestibule in three difference planes of space, one

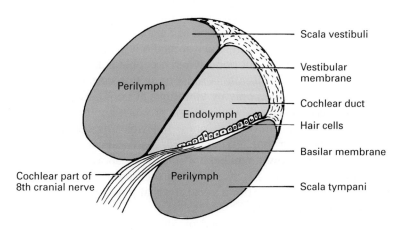

Fig. 6.3 Section through the cochlea.

vertical, one horizontal and one transverse. They contain peri-lymph. When the position of the head is altered the movement of the endolymph stimulates special cells with hair-like processes situated at the end of each canal. This information helps in the maintenance of posture, though in daylight it is mainly the respon-sibility of the eyes to supply information about the position of the head in space. Overstimulation of the fluid in the semicircular canals causes giddiness.

The *membranous labyrinth* is contained within the bony labyrinth although it is much smaller. It includes the utricle, the saccule, the semicircular ducts and the cochlear duct.

The *utricle* and *saccule* are two small sacs in the vestibule which communicate with each other through a connecting tube. They contain patches of sensory hair cells which are stimulated by the action of gravity on small crystals (otoliths) which adhere to them.

The *semicircular ducts* are similar in shape to the semicircular canals and lie within them but are only a quarter of the diameter. They contain endolymph.

The *cochlear duct* is a spiral tube within the bony canal of the cochlea and lying along its outer wall. Its roof is formed by the vestibular membrane and its floor by the basilar membrane and its outer wall by the bony wall of the cochlea.

THE MECHANISM OF HEARING

The sound wave is a wave of *vibration* of the air set up by the vibration of an object. The vibration of a fiddle-string or the vocal cord, for instance, sets up vibration of the air in contact with it and produces waves of vibration which spread out in all directions, like the ripples set up in a still pond when a pebble is thrown into it.

To produce sound, vibrations must be within certain rates. The human ear is stimulated only by vibrations at rates between 30 and 30 000 per second. The slow vibrations produce low notes and the quick vibrations produce high ones. It is for this reason that a man's voice is lower than a woman's, for his vocal cords are longer and vibrate more slowly while the woman's are shorter and vibrate more quickly. It is the rapid growth in the larynx and the conse-quent lengthening of the boy's vocal cords at puberty which causes the breaking of the voice.

Sound waves travel at a speed of 340 metres per second. They travel much more slowly than light rays, hence the flash of light-

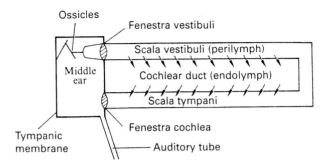

Fig. 6.4   Diagram to show how vibrations pass through the inner ear.

ning is seen before thunder is heard, and the farther away the storm the longer the interval between the two.

Sound waves normally are carried by air, but they also pass through solid bodies; in fact, a solid carries sound more readily than air. Thus, putting an ear to the ground, it is possible to hear footsteps at a greater distance than through the air. It is, however, normally air with which the ear is in contact.

Hearing is due to the sound waves making the tympanic membrane vibrate, which sets the ossicles and fenestra vestibuli vibrating, which in turn causes vibration of the perilymph (Fig. 6.4). As fluid is incompressible, the perilymph can vibrate only if the fenestra cochlea is able to bulge outwards as the fenestra vestibuli bulges inwards. Hence the need for two windows in the internal ear. Vibration of the perilymph gives rise to vibration of the endolymph, and this affects the little hairs which jut into it and stimulates the endings of the vestibulo-cochlear nerve in the membranous cochlea. The nerve carries the stimulus to the centre of hearing in the temporal lobe of the brain, where it is appreciated and interpreted.

The appreciation of sound will result from any stimulus brought by the auditory nerve to the centre of hearing, but the meaning given to the sound will depend on previous experience and the power of reasoning.

Chapter 7
The Eye

Situated in the orbit, the eye is the organ of sight and, as such, its main function is to focus light onto the retina. The retina is composed of nervous tissue which sends signals generated by light to the brain. This chapter will look at how the structure of the eye can achieve this and how the eye, which is a very sensitive organ, is protected and in order to maintain its function.

THE STRUCTURE OF THE EYE

The eye is spherical in shape and is embedded in fat. It has three coats: an outer fibrous coat, a vascular, pigmented coat and an inner nervous coat (Fig. 7.1).

The outer *fibrous coat* has two parts. The posterior part is opaque and is called the *sclera*; it is a firm membrane which preserves the shape of the eyeball. Its external surface is white and forms the

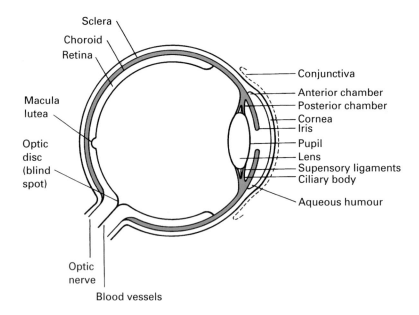

Fig. 7.1 The eye.

white of the eye. The anterior part of the *sclera* is covered with conjunctiva, which is reflected on to it from the inner side of the eyelid and is continuous with the corneal epithelium covering the cornea. The *cornea* is the anterior part of the fibrous coat. It projects a little from the surface of the eye and is transparent, allowing light rays to enter the eye and bending them to focus on the retina (refraction).

The *vascular, pigmented coat* has three parts. The *choroid* lines all but the front part of the eye. It is dark brown in colour and supplies blood to the other layers of the eye, particularly the retina. The *ciliary body* is a thickened part of the middle coat containing muscular and glandular tissue. The *ciliary muscles* control the shape of the lens, enabling it to focus light rays from near or far away as required. They are known as the muscles of *accommodation*. The *ciliary glands* produce a watery fluid, the *aqueous humour*, which fills the eye in front of the lens and passes into the veins through small openings in the angle between the iris and the cornea. The *iris* is the coloured part of the eye. It lies between the cornea and the lens and divides the space between them into anterior and posterior chambers. The iris contains muscular tissue arranged in circular and in radiating fibres; the circular fibres contract the pupil and the radiating fibres dilate it. There is a circular opening in the centre called the *pupil* which is contracted in bright light to prevent too much light entering the eye and dilated in poor light to allow as much light as possible to reach the retina (Fig. 7.2).

The inner lining of the eye is called the *retina*. It is a delicate membrane adapted for the reception of light rays and contains many nerve cells and fibres. It is made of *rods* and *cones* (Fig. 7.3) which are thought to have separate functions. The cones are more numerous in the centre of the eye and are responsible for detailed vision and colour perception; the rods are more numerous around the outer edge of the retina and are sensitive to the movements of

Circular fibres   Radiating fibres
    constrict   dilate

Fig. 7.2   The pupil.

Fig. 7.3   Layer of rods and cones on the outside of the retina.

objects within the field of vision. They contain a pigment called *visual purple* for the synthesis of which vitamin A is required; lack of vitamin A in the diet may result in *night blindness*. Near the centre of the back of the retina there is an oval yellowish area called the *macula lutea* where only cones are present and which is the area where vision is most perfect. About 3 mm to the nasal side of the macula lutea the optic nerve leaves the eye; this area is called the *optic disc* and because it is insensitive to light it is called the blind spot. An object can only be opposite the blind spot in one eye at a time. To mark a blind spot, mark a piece of paper as shown below:

X   ●

Now shut the left eye, fix the right eye steadily on the cross, and move the paper slowly backwards and forwards on eye level. At a certain point the dot will disappear because it is opposite the blind spot.

The contents of the eye are listed in Box 7.1.

The contents of the eye

- The aqueous humour
- The vitreous humour
- The lens

Box 7.1

The *aqueous humour* has already been described on page 93.

The *vitreous humour* is a colourless, transparent jelly-like substance which maintains the shape of the eyeball.

The *lens* is situated immediately behind the iris. It is a transparent, biconvex body enclosed in a transparent, elastic capsule from which ligaments pass to the ciliary body. These *suspensory ligaments* hold the lens in position and are the means by which the ciliary muscles exert pull on the lens, altering its shape for near or far vision.

THE MECHANISM OF SIGHT

As light rays pass through the transparent cornea, aqueous humour and lens they are bent, a process known as *refraction*. This makes it possible for light from a large area to be focused on a small area of the retina. Parallel light rays striking a convex lens are bent towards a focal point on the retina (Fig. 7.4). If the object is less than seven metres away the curvature of the lens must be increased to enable the focus to be on the retina. This is called *accommodation* (Fig. 7.5). Far vision can be achieved with the lens in its normal resting position.

Some people are naturally *short-sighted*. This is due to the fact that their eyes are too long, so that the retina is farther from the lens than it should be and the focusing point lies in front of it. In this case the near object can be seen with the flatter lens of the eye at

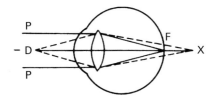

Fig. 7.4   The normal eye at rest. P, Parallel rays from a distant object focused by the lens on the retina (F). Dotted lines show the diverging rays from a near object (D) of which the focus falls behind the retina at X.

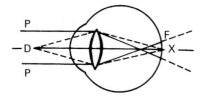

Fig. 7.5   The normal eye during accommodation. Rays (dotted lines) from a near object (D) are focused on the retina at point X by the more curved lens. P, Parallel rays from a distant object are now focused in front of the retina (F).

rest (normally used for far sight), but for far sight concave glasses are necessary to throw the focus point farther back.

Others are naturally *long-sighted*. This is due to the fact that their eyes are too short, so that the retina is too close to the lens and the focusing point lies behind it. In this case the far object can be seen by using the thicker, more curved lens used by the normal person for near sight, as this will bend the light rays more and bring the focusing point forward; for near sight, since they are already using the more curved lens, they must be provided with convex glasses to bend the light rays from the near object still more acutely (Figs. 7.6 and 7.7).

The eye is moved within the orbit by six orbital muscles which are ribbon-like and are attached to the sclera. These muscles pull

Fig. 7.6   A convex lens, showing how parallel light rays are brought to focus at F.

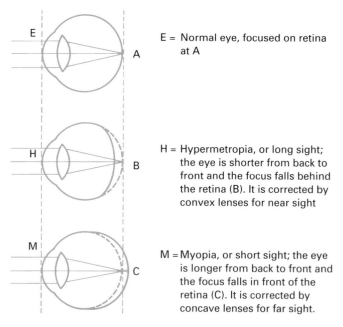

E = Normal eye, focused on retina at A

H = Hypermetropia, or long sight; the eye is shorter from back to front and the focus falls behind the retina (B). It is corrected by convex lenses for near sight

M = Myopia, or short sight; the eye is longer from back to front and the focus falls in front of the retina (C). It is corrected by concave lenses for far sight.

Fig. 7.7   The normal eye, hypermetropia and myopia.

on the eyes and co-ordinate their movement so that both eyes focus on the same object. A weakness in one or more muscles will cause one eye to deviate, a condition that is commonly known as a *squint*.

PROTECTION OF THE EYES

The eyes are very delicate organs and are protected by the eyebrows, the eyelids and the lacrimal apparatus, as well as by the bony orbits in which they lie embedded in fatty tissue.

The overhanging *brows* protect them from injury and excessive light, while the hairs entangle sweat and prevent it from running into the eyes.

The *eyelids* consist of a plate of fibrous tissue covered by skin and lined with mucous membrane. The edges of the lids are provided with hairs, the eyelashes, which keep out dust, insects and too much light. The transparent mucous membrane which lines the lids is reflected over the front of the eyeball and is called the *conjunctiva*. This results in the formation of upper and lower conjunctival sacs under the upper and lower lids respectively. Dust and bacteria tend to stick to the moist surface of this membrane, and to keep it clean it is constantly washed by the lacrimal apparatus.

The *lacrimal apparatus* (Fig. 7.8) consists of:

Fig. 7.8 The lacrimal apparatus.

1.   The *lacrimal gland,* lying over the eye at the outer side and secreting lacrimal fluid into the conjunctival sac.

2.   Two fine canals, called *lacrimal canaliculi,* leading from the inner angle of the lids to the lacrimal sac.

3.   The *lacrimal sac,* which lies at the inner angle of the eyelids in the groove on the lacrimal bone.

4.   The *naso-lacrimal duct,* which runs from the lacrimal sac down to the nose.

The opening into the canaliculi can be seen at the inner angle of the eyelids and is called the *punctum.*

The fluid secreted by the lacrimal glands washes over the eyeball and is swept up by the blinking action of the eyelids. The muscles which cause the blink press on the lacrimal sac and contract it so that as they relax the sac expands and sucks the fluid from the edges of the lids along the fine canals into the sac; from this it runs by gravity down into the nose. Thus the window which admits light into the eye is constantly irrigated by a gentle stream of fluid, which keeps it clean and washes away germs and harmful substances. The fluid is composed of water, salts and an antibacterial substance called *lysozyme.*

Chapter 8
Endocrine Glands

The endocrine glands are organs producing secretions called *hormones* which are poured directly into the blood stream from the glandular cells. It is for this reason that they are also known as *ductless glands*. This chapter is mainly concerned with the overall control of the endocrine system, which is composed of all the endocrine glands, by the pituitary in the brain. The function of the thyroid and parathyroid glands will be considered in some detail along with the adrenal glands, although the gonads, which are also endocrine glands, are considered in Chapter 27.

Hormones are organic compounds manufactured by the glands from substances carried in the blood. They are mainly protein derivatives, but some are steroids. They are then carried by the blood stream to other parts of the body where they have a specific effect.

The endocrine glands have been investigated and are described as though they are separate entities but in fact their functions are closely interrelated. Initially the functions of the hormones were deduced by observing the effects of disease, destruction or overgrowth of glands. In recent years hormones have been isolated, obtained in pure form, analysed and in some cases successfully synthesized. The glands secrete their hormones continuously but the quantity of secretion can be increased or decreased according to body needs. Control of hormone secretion occurs in several different ways (Box 8.1).

Control of hormone secretion

- Nerve cells produce chemical substances which are carried to the gland and cause secretion
- The gland responds to impulses from the autonomic nervous system
- One gland produces a hormone which affects a second gland. The second gland produces its hormone, which influences secretion of the first gland. This is called a 'feedback' mechanism
- The gland responds to blood levels of substances other than hormones

Box 8.1

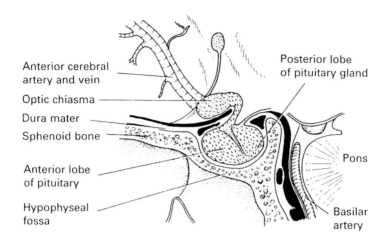

Fig. 8.1   The pituitary gland, shown in section, lying in the sella turcica or hypophyseal fossa. Note the stalk joining it to the base of the brain.

THE HYPOPHYSIS

The hypophysis, or pituitary gland, lies in the hypophyseal fossa of the sphenoid bone in the base of the skull. It is attached by a neural stalk to the *optic chiasma* at the base of the brain (Fig. 8.1).

The gland consists of an *anterior lobe*, or *adenohypophysis* and a *posterior*, or *neural*, lobe. The anterior lobe is an endocrine gland in the true sense, while the posterior lobe is derived from the brain and consists of nervous tissue; it is connected directly to the *hypothalamus*. The two lobes are essentially two different endocrine glands and are commonly called the anterior and posterior pituitary glands.

The anterior lobe of the hypophysis is sometimes referred to as the 'master gland' of the endocrine system because it has an important influence in regulating the function of other glands. However, the glands really act in concert, one becoming active if the others are not producing sufficiently and output of hormones being decreased when other glands are active.

The anterior lobe produces a number of hormones (Box 8.2):

1. *Thyroid-stimulating hormone* (*TSH* or *thyrotrophic hormone*) influences all aspects of thyroid gland function including stimulation of the accumulation of iodine in the gland for conversion into thyroid hormones, manufacture of the hormones and their

---

Hormones of the pituitary

Anterior lobe
- Thyroid-stimulating hormone (TSH)
- Adrenocorticotrophic hormone (ACTH)
- Somatotropin
- Follicle-stimulating hormone (FSH)
- Leuteinizing hormone (LH)
- Prolactin

Posterior lobe
- Oxytocin
- Antidiuretic hormone (ADH)

Box 8.2

---

release into the circulatory system. Through its action on the thyroid gland TSH is involved in the regulation of metabolic rate, breakdown of fat and the increase of the water content of certain tissues.

2. *Adrenocorticotrophic hormone (ACTH)* regulates the development, maintenance and secretion of the cortex of the suprarenal glands. General metabolic effects of this hormone include mobilization of fats, production of hypoglycaemia and increase in muscle glycogen. It is also involved in body resistance to stress.

3. *Somatotrophic (growth) hormone* exerts its influence mainly on the hard tissues of the body, though there is some effect on the soft tissues. The hormone increases the rate of growth and maintains size once maturity has been reached. It controls the rate of growth in epiphyses and other ossification centres of the skeletal system. Oversecretion of this hormone causes overgrowth of the long bones in children (gigantism) and acromegaly in adults. In acromegaly the bones cannot increase in length because the epiphyseal plates have closed so the bones become thicker and coarser; the lower jaw, hands and feet are particularly affected. Undersecretion of the growth hormone produces dwarfism. People who are very short or very tall, due to undersecretion or oversecretion of this hormone, are usually of normal intelligence, unlike the person who suffers from undersecretion of the thyroid gland. Metabolic effects of this hormone include an increase in the conversion of carbohydrates to amino acids and in the uptake of amino acids by the cells; mobilization of fat from

the storage areas and increased fat metabolism and an increased blood sugar level.

4.   *Follicle-stimulating hormone* (*FSH*) controls the maturation of ovarian follicles in the female and the production of sperm in the male.

5.   *Luteinizing hormone* (*LH*) causes changes in the female which lead to the formation of the corpus luteum; it also helps to prepare the breasts for the secretion of milk. In the male a comparable hormone is called *interstitial cell stimulating hormone* (*ICSH*) which acts on the testes and controls the secretion of the male sex hormone testosterone.

6.   *Lactogenic hormone* (*prolactin*) is one of several hormones involved in the production of milk by the breasts and appears to function only in females.

The posterior lobe of the hypophysis releases two hormones (Box 8.2), but it is important to note that these are produced in the hypophalamus and only stored in, and released from, the posterior lobe.

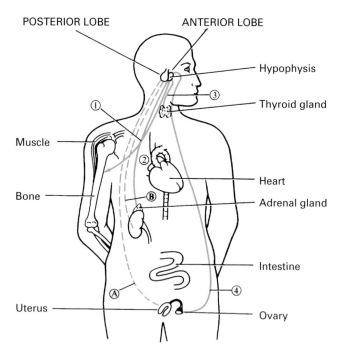

Fig 8.2 Diagram to show the effects of the pituitary hormones. The numbers and letters refer to those in Table 8.1.

1.   *Oxytocin* exerts its effects mainly on the smooth muscle of the pregnant uterus and the cells around the ducts of the breasts, although it does also promote a generalized contraction of unstriped muscle throughout the body.

2.   *Antidiuretic hormone* (*ADH* or *vasopressin*) causes an increase in the reabsorption of water by the kidney tubules so that less urine is excreted. Undersecretion of ADH causes less water to be reabsorbed and excessive amounts of very dilute urine are excreted—a condition known as *diabetes insipidus*. The hormone also causes a certain amount of vasoconstriction with consequent rise in blood pressure, but in man this occurs mainly in the coronary vessels.

The effects of the pituitary hormones are summarized in Fig. 8.2 and Table 8.1.

THE THYROID GLAND

The thyroid gland is situated at the front and sides of the neck, opposite the lower cervical and first thoracic vertebrae. It consists of two lobes, one on either side, joined by a narrower portion called the isthmus which crosses in front of the trachea just below the larynx (Fig. 8.3). The gland is formed of a number of closed *follicles* containing a yellow semi-fluid material called *colloid* (Fig. 8.4).

The cells produce two hormones called *thyroxine* and tri-iodothyronine (known as T_4 and T_3, respectively), which may be released directly into the blood stream if it is required or may be linked to the protein substance, *thyroglobulin*, and stored in the colloid. The amino acid *tyrosine* and the mineral iodine are both essential for the formation of thyroid hormones. Tyrosine in the thyroid hormones binds iodide. Thyroxine (T_4) binds four iodide ions and tri-iodothyronine (T_3) binds three iodide ions.

The function of thyroxine is to regulate metabolism in the tissues. Together with growth hormone it ensures proper development of the brain; it increases urine production, protein breakdown and the uptake of glucose by the cells. Tri-iodothyronine, has a more immediate, though similar effect to thyroxine.

Undersecretion of thyroid hormones in a child produces the condition of *thyroid cretinism* which if untreated results in a mentally retarded dwarf. In an adult, undersecretion produces the condition known as *myxoedema*. In both these conditions the skin is dry and coarse, and the hair dry, coarse and lank. The metabolic

Table 8.1 Action of Pituitary Hormones

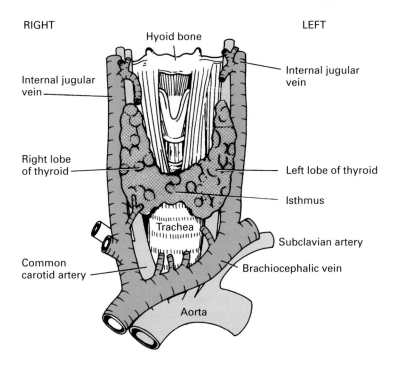

Fig 8.3 The thyroid gland.

rate is lowered so the patient is obese, has a low body temperature and feels the cold. Oversecretion of the hormones produces *thyrotoxicosis*, a condition in which there is an increase in the metabolic rate. The individual is anxious and nervous and has a fast pulse rate. The skin is fine and moist, the patient feels the heat and loses weight in spite of having a good appetite. He may or may not have protruding eyeballs, a condition known as exophthalmos.

Any enlargement of the gland is known as *goitre* and this may be present with or without hyperthyroidism. The presence of a goitre may cause pressure on the trachea or on the recurrent laryngeal nerve causing hoarseness.

THE PARATHYROID GLANDS

The parathyroid glands usually lie between the posterior borders of the lobes of the thyroid gland and its capsule. They are about the size of a pea and there are usually four—two behind each lobe. This number occasionally varies.

Fig 8.4   Thyroid follicles.

The glands produce a hormone called *parathormone* which regulates the distribution and metabolism of calcium and phosphorus in the body. Oversecretion causes calcium to be moved from the bones into the blood, from which it is excreted in the urine. The bones become porous and brittle and the increased level of blood calcium may cause the formation of kidney stones (renal calculi). Undersecretion of the hormone causes low blood calcium levels resulting in the muscular rigidity and spasm seen in the condition known as *tetany*. This complication may occasionally be seen after surgery on the thyroid gland in which the parathyroids have inadvertently been removed. Calcium is given to correct the condition.

THE ADRENAL GLANDS

The adrenal glands lie one above and in front of the upper end of each kidney, behind the peritoneum. They are surrounded by areolar tissue containing a considerable amount of fat. Each gland consists of two quite separate endocrine glands: the outer part called the cortex and the inner part called the medulla (Box 8.3).

Cortex

The cortex is subdivided into three zones: the outer zone produces the mineralocorticoids, the central zone produces the glucocorticoids, the inner zone produces sex hormones.

Adrenal hormones

Cortex
- Mineralocorticoids
- Glucocorticoids
- Sex hormones

Medulla
- Adrenaline
- Noradrenaline

Box 8.3

Mineralocotricoids are the steroids that regulate electrolyte metabolism. *Aldosterone* is an important mineralocorticoid and it has a regulatory effect on the relative concentrations of minerals in the body fluids, particularly sodium and potassium. It therefore also affects the water content of the tissues. When there is a deficiency of this hormone there is increased excretion of sodium and chloride ions, too much water is lost from the body in the urine and there is a corresponding fall in the concentrations of sodium, chloride and bicarbonate in the blood, resulting in a lowered pH (or acidosis).

Glucocorticoids are important in carbohydrate metabolism. They increase the conversion of protein to glycogen for storage by the liver (gluconeogenesis) and decrease the utilization of glucose by the cells, thus increasing the blood sugar level. *Cortisone*, *cortisol* (hydrocortisone) and *corticosterone* are the primary glucocorticoids and they may be given to people who have chronic inflammatory and allergic responses, such as in rheumatoid arthritis, because the steroids stimulate the anti-inflammatory and repair mechanisms of the body. Chronic stress of any kind produces increased production of the glucocorticoids which in turn help the body to resist stress; however, continuing high levels of glucocorticoids may increase the likelihood of ulcer formation, increase the blood pressure and lower the body's resistance to infection by causing atrophy of lymphatic tissue.

Sex hormones are the *androgens* (male hormones) and *oestrogens* (female hormones) and the effects are similar to those of the hormones produced by the testes and the ovaries. Adrenal androgens and oestrogens are produced in minute quantities in both sexes and affect the development and function of the reproductive organs and the physical and temperamental characteristics of men and women. Their effects become noticeable when there is oversecretion; for example, an adrenal tumour in a woman may produce secondary masculine characteristics such as the growth of hair on the face and deepening of the voice.

Undersecretion of the cortical hormones results in a condition known as Addison's disease which produces symptoms of anaemia, muscular weakness, low blood pressure, low blood sugar level, and bronzing of the skin and mucous membranes.

Oversecretion of the cortical hormones results in several disorders. Cushing's disease is due mainly to overproduction of the glucocorticoids. There is excess fatty tissue on the trunk and face but not the limbs and sodium is retained so there is oedema, increased plasma volume and a raised pH (alkalosis).

Medulla

The medulla of the adrenal glands produces two hormones, *adrenaline* and *noradrenaline*, which have effects similar to that obtained by stimulation of the sympathetic nervous system. Because control of the medulla is by way of preganglionic neurones of the sympathetic nervous system, without interruption, response to stimulation is very rapid and creates a set of conditions in which the body is prepared for 'fight or flight'. *Adrenaline* increases the strength and rate of the heart beat; causes dilation of arteries supplying the heart and skeletal muscles but constriction of other arteries; increases rate and depth of respiration; increases blood sugar level by promoting breakdown of liver glycogen and stimulates the general metabolic activity of the cells.

THE THYMUS GLAND

The thymus gland is an organ which lies in the lower part of the neck and chest between the lungs over the heart. It varies in size with age and grows until the child is two years of age; it then shrinks so that in the adult only fibrous remnants are found. When fully developed it is greyish-pink in colour and consists of two or three lobes. Its structure resembles that of a lymph node and it is similarly associated with antibody formation. No hormone has yet been isolated from it and it may well belong to the circulatory rather than the endocrine system.

THE PINEAL BODY

The pineal body is a small reddish body about the size of a cherry stone, lying behind the third ventricle of the brain. Its function is unknown. In later life it becomes calcified and acts as a useful landmark in X-rays of the skull.

THE GONADS

The gonads, which are responsible for the production of ova in females and sperm in males, are under the control of FSH and LH. Their function, which also includes endocrine functions, is discussed in Chapter 27.

THE PANCREAS

The pancreas has an endocrine function. It contains an area known as the *islets of Langerhans* which produces glucagon and insulin from the α and β cells, respectively. These hormones influence glucose metabolism, which is discussed in Chapters 21 and 23.

SECTION III: POSTURE AND MOVEMENT

Section Introduction

Posture and movement, which are controlled by the central nervous system, are achieved by the combined actions of the skeletal and muscular systems. The skeletal system is composed of the bones which make up the skeleton, and the muscular system is composed of the skeletal, or voluntary, muscles which are attached to the skeleton.

There are different types of bone, and these, in turn, are capable of different types and degrees of movement. Movement between two bones is called 'articulation' and the system of joints whereby this is achieved is described.

Skeletal muscle is a unique type of tissue and its gross structure and organization into groups of muscles, as well as the mechanism whereby movement is achieved, are all considered here. The utilization of chemical energy, in the form of ATP, which is produced in metabolism, is introduced here.

Chapter 9
Development and Types of Bone

The skeletal system is made up of just over 200 bones, joined together to provide a strong, movable, living framework for the body. The system has four main functions. It supports and protects the surrounding soft tissues and vital organs; it assists in body movement by giving attachment to muscles and providing leverage at joints; it manufactures blood cells in the red bone marrow and it provides storage for mineral salts, particularly phosphorus and calcium. This chapter looks at the developmental processes undergone by bone in the production and maintenance of the human skeleton. The ways in which different types of bone can be described anatomically are also outlined.

Cartilage provides the environment in which bone develops. Development is from spindle-shaped cells, called *osteoblasts*, which are found underneath the tough sheath of fibrous tissue which covers the bone called the periosteum, and in cavities within the bone. These cells are able to convert the soluble salts of calcium and magnesium in the blood, such as calcium chloride, into insoluble calcium salts, chiefly calcium phosphate. A second set of cells is also present which can change the insoluble calcium phosphate into soluble calcium salts which are carried away by the blood. These cells are capable of causing the absorption of any unwanted bone and are known as *osteoclasts*. Both types of cell are active during the growth period, osteoblasts producing bone and osteoclasts removing it to maintain the form and proportion of the bone. For example, osteoblasts build up bone on the surface of a hollow bone, while osteoclasts absorb bone on the inner side to enlarge the cavity and prevent the bone from becoming too heavy.

OSSIFICATION

There are two types of ossification (Box 9.1).

BONE GROWTH AND REPAIR

Plentiful supplies of calcium and phosphorus are essential in the diet of the pregnant and nursing mother, the growing child and

---

Types of ossification

Intramembranous ossification
- A process in which dense connective tissue is replaced by deposits of calcium salts, forming bone
- The bones of the skull form by this process

Intracartilaginous ossification
- A process in which cartilaginous structures are replaced with bone
- Most bones form by this process

Box 9.1

---

where bone repair is taking place after injury or disease. Calcium is present in milk and eggs and in green vegetables, and phosphorus in meat, fish and the yolks of eggs. Vitamin D is necessary so that calcium and phosphorus may be absorbed from the intestine and fully utilized within the body; lack of vitamin D causes rickets in children and osteomalacia in adults. Both these conditions cause soft bones, which bend under the weight of the body and various deformities of the weight-bearing bones may result. Vitamin D is found in animal fat and fish oils and in margarine to which it has been added.

The human body can also manufacture the vitamin. The action of the ultra-voilet rays of the sun converts *ergosterol* in the skin into vitamin D.

Vitamin C is important in bone growth as it plays a part in laying down *collagen*, which is the main constituent of connective tissue. It is found in fresh fruit, particularly citrus fruit, blackcurrants, green vegetables, tomatoes and potatoes.

Growth and development of bone is also affected by both *exercise* and *rest*. Exercise causes an increased blood supply to the muscles and therefore to the underlying bones; because the blood carries necessary building materials exercise results in increased growth and recognition of this fact is responsible for increased attention to physical exercise in schools. The pull of developing muscles is an important factor in determining the shape of the bones and posture may also affect shape by altering stresses on bones. The importance of rest is interesting. The bones of a child are comparatively elastic so that at the end of a day of standing and running about there is an appreciable loss of height. During rest the bones recover their full length, so a mid-day rest and a long night will affect growth.

A further factor controlling growth is the secretions of certain ductless glands. This was mentioned in Chapter 8.

TYPES OF BONE TISSUE

Bone tissue is of two types, compact and spongy.

Compact bone appears to be solid but when it is examined under a microscope it is found to consist of *Haversian systems* (Box 9.2; Fig. 9.1).

Components of the Haversian system

- A central canal, called a **Haversian canal** which contains blood vessels, nerves and lymphatics
- Plates of bone, called *lamellae* arranged round the central canal.
- Between the lamellae are spaces called *lacunae* which contain bone cells, called *osteocytes*, and lymph.
- Fine channels, called *canaliculi*, between the lacunae and the central canal carry lymph which brings nutrients and oxygen to the osteocytes.

Box 9.2

Between the Haversian systems there are tiny circular plates of bone called *interstitial lamellae*.

Spongy bone is hard like all bone but has a spongy appearance to the naked eye. When examined under a microscope the Haversian

Fig. 9.1   The structure of compact bone.

canals are seen to be much larger and there are fewer lamellae. The spaces in spongy bone are filled with *red bone marrow* which consists of fat and blood cells and in which red blood cells are made.

There are three types of bone (Box 9.3).

Types of bone

- Long bones
- Flat bones
- Irregular bones

Box 9.3

Long Bones

A long bone consists of a shaft and two extremities (Fig. 9.2). The shaft has an outer layer of compact bone surrounding a central cavity called the *medullary canal* which contains yellow bone marrow. Yellow bone marrow, like red bone marrow, consists of fat and blood cells, but it is not so rich in blood supply or in red blood cells. The extremities consist of a mass of spongy bone containing red bone marrow covered by a thin layer of compact bone. The bone is covered by a tough sheath of fibrous tissue called the *periosteum*; this is richly supplied with blood vessels which pass into the bone to nourish it.

Three distinct sets of blood vessels supply the long bones:

1. Innumerable tiny arteries run into the compact bone to supply Haversian canals and systems.
2. Many larger arteries pierce the compact bone to supply the spongy bone and red bone marrow. The openings by which these vessels enter the bone can be seen quite easily by the naked eye.
3. One or two large arteries supply the medullary canal. These are known as nutrient arteries and they pass through a large opening called the nutrient foramen, which runs obliquely through the shaft to the medullary canal.

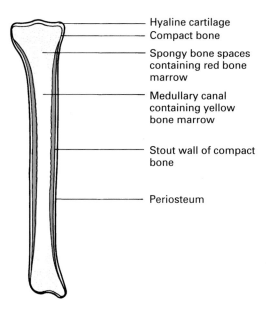

Hyaline cartilage
Compact bone
Spongy bone spaces containing red bone marrow
Medullary canal containing yellow bone marrow
Stout wall of compact bone
Periosteum

Fig. 9.2 The structure of a long bone.

These three sets of vessels are linked by fine branch arteries within the bone.

The periosteum nourishes the underlying bone through its blood vessels. If it is torn off the underlying bone will die, but if the bone is destroyed by disease and the periosteum remains healthy it is able to build up new bone. The periosteum is also responsible for growth in the thickness of the bone through the action of osteoblasts, adjacent to the bone surface, which are able to lay down fresh bone. The periosteum is protective in function and also gives attachment to tendons of muscles. It is not present on the joint surfaces of a bone where it is replaced by hyaline cartilage, called articular cartilage, which provides a smooth surface so that joint movements can occur without friction.

Development of Long Bones

Long bones develop from three centres of ossification, one in the shaft and one or more at either extremity (Fig. 9.3). The centre of ossification in the shaft is known as the *diaphysis*. Those at the extremities are called *epiphyses* and begin to develop after birth. From these centres ossification gradually spreads throughout the extremity (Figs. 9.4 and 9.5) and is well developed by 12 years of

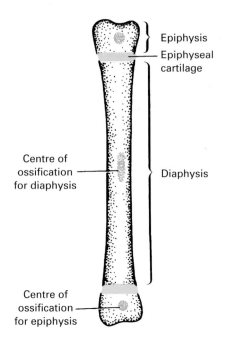

Fig. 9.3   The development of a long bone (growth in length occurs at the epiphyses).

Fig. 9.4   X-ray of the hand of a 1-year-old infant. The shafts of the small long bones of the fingers and the thumbs are well formed, but note the gaps (1) between them and the knuckles. Note also that only two of the eight bones in the wrist (2) have begun to ossify.

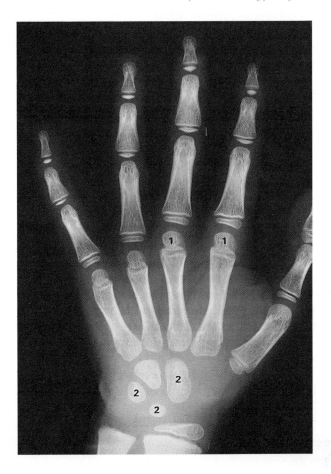

Fig. 9.5 X-ray of the hand of a 6-year-old child. Note (1) the development of the separate extremities, i.e. the epiphyses, of the long bones of the hand, and (2) the ossification that has occurred in the wrist or carpal bones. Only certain bones are marked for comparison.

age, though there is still a line of cartilage left between the shaft and the extremity.

The line of cartilage between epiphysis and diaphysis is known as the *epiphyseal cartilage* and it is from this cartilage that growth in length occurs. It is the shaft that grows longer and fresh bone is constantly produced at either end of it by the epiphyseal cartilage. When full growth has been attained, between the ages of 18 and 25 years, the line of cartilage turns to bone and no further growth can occur. The timing of this event varies from one bone to another and

Fig. 9.6. The structure of a flat bone.

from one end of a bone to the opposite end of the same bone. For example, the lower epiphysis of the humerus unites at about the 18th year, but the upper epiphysis does not join the shaft until about two years later.

Flat Bones

Flat bones consist of two stout layers of compact bone joined by a layer of spongy bone (Fig. 9.6). These bones are also covered by periosteum from which two sets of blood vessels pass into the bone to supply the compact and spongy bone respectively. They are found in the head, trunk, shoulder girdle and pelvic girdle, where they give protection to the delicate organs underlying them and provide attachment for the powerful muscles required to control the freely movable shoulder and hip joints. The layers of compact bone in the flat bones of the skull are referred to as the *outer* and *inner tables* and the spongy bone is called the *diploë*. The outer table is thick and strong so that it is not readily broken. In some areas the spongy bone is absorbed leaving air-filled cavities between the two tables which are called *sinuses*.

Irregular Bones

These consist of a mass of spongy bone covered by a thin layer of compact bone (Fig. 9.7). They are covered with periosteum, except on the articular surfaces, and as with flat bones this protective

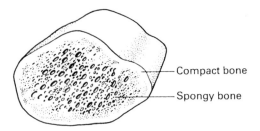

Fig. 9.7 The structure of an irregular bone.

Surface irregularities of bone

Articular projections and depressions
These enter into the formation of joints and are smooth

Articular projections
- Head—round, like a sphere or a disc
- Condoyle—rounded but oval in outline, like the typical knuckle bone

Articular depressions
- Socket—a depression into which another organ fits
- Fossa—a shallow depression

Non-articular projections and depressions
These give attachment to muscles or ligaments and are rough

Non-articular projections
- Process—a rough projection for muscle attachment
- Spine—a pointed rough projection
- Tuberosity—a broad rough projection
- Trochanter—a broad rough projection
- Tubercule—a small tuberosity
- Crest—a long rough narrow projecting surface

Non-articular depressions
- Fossa—a notch in the bone
- Groove—a long narrow depression

Other terms applied to bones
- Foramen—an opening in the bone
- Sinus—a hollow cavity in the bone

Box 9.4

---

covering provides two sets of blood vessels to supply both compact and spongy bone. They are found in the spinal column and the middle ear and also at the ankle and wrist, although these latter are sometimes called short bones.

SURFACE IRREGULARITIES

The surfaces of all bones are very irregular and show numbers of *depressions* and *projections*. These may be divided according to function (Box 9.4).

All the rough non-articular projections give attachment to muscles; the stronger the muscle and the more it is used, the larger and rougher the projection, providing a greater surface for muscle attachment. In a paralysed limb the processes either fail to develop, or atrophy, according to age.

Chapter 10
Bones of the Head and Trunk

The student is advised that it is not possible to study adequately the bones of the skeleton without access to an entire skeleton and to disarticulated bones which can be handled and examined closely.

The skeleton (Fig. 10.1) can be divided into groups of bones (Box 10.1).

The skeleton

- The bones of the head
- The bones of the trunk
- The bones of the upper limb and shoulder girdle
- The bones of the lower limb and pelvic girdle

Box 10.1

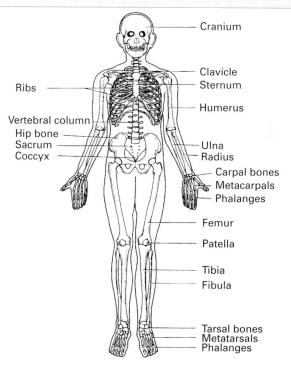

Fig. 10.1   The skeleton (male).

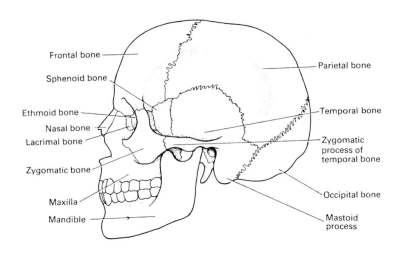

Fig. 10.2   The bones of the head.

The bones of the head and trunk are also known as the *axial skeleton*, the main support of the body, while the bones of the extremities may be known as the *appendicular skeleton*.

THE BONES OF THE HEAD

For purposes of description the skull (Fig. 10.2) may be divided into:

1.   The bones of the cranium
2.   The bones of the face

The Bones of the Cranium

The cranium is a box-like cavity which contains and protects the brain. It has a dome-shaped roof called the *calvaria* or skull cap and its floor is known as the *base of the skull*. The cranium consists of fifteen bones (Box 10.2).

The *frontal bone* is a large flat bone forming the forehead and most of the roof of the orbit. There are rounded prominences, called the *frontal tuberosities*, one on each side of the midline, which vary in size from one individual to another and which together form the forehead. The bone contains two irregular cavities called the *frontal sinuses* which lie one over each orbit and which open into the nasal

---

The bones of the cranium

- One frontal bone
- Two parietal bones
- One occipital bone
- Two temporal bones
- One ethmoid bone
- One sphenoid bone
- Two inferior nasal conchae
- Two lacrimal bones
- Two nasal bones
- One vomer

Box 10.2

---

cavity. The sinuses contain air and are lined with mucous membrane which is continuous with the mucous membrane lining the respiratory tract. They add resonance to the voice and they serve to lighten the skull, but the mucous membrane may become infected, causing a condition known as *sinusitis*.

The *parietal bones* form the sides and roof of the cranium; they articulate with the frontal bone, the occipital bone and with each other to form the *sutures* or joints of the cranium (see Chapter 12). On the internal surface are small grooves to carry the blood vessels supplying the brain and the impression of the folds or convolutions of the surface of the brain can be seen. At birth there are membranous gaps in the skull at the angles of the parietal bone which are called *fontanelles* (see Chapter 12).

The *occipital bone* (Fig. 10.3) forms the back of the skull. It carries a marked prominence, the *external occipital protuberance* which gives attachment to muscles. Below this there is a large oval opening, known as the *foramen magnum*, through which the cranial cavity communicates with the vertebral canal. On either side of the foramen are two smooth oval processes called the *occipital condyles* for articulation with the first cervical vertebra. This joint allows the nodding movement of the head.

The *temporal bones* (Fig. 10.4) are situated at the sides and base of the skull. Each consists of four parts:

1.   The *squamous part* forms the anterior and upper part of the bone and is thin and flat. A long arched process, called the *zygoma* or zygomatic process projects forward from the lower portion.

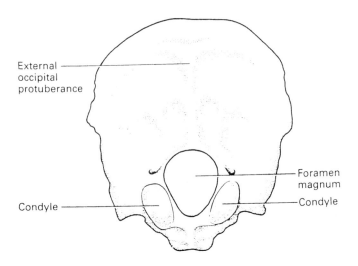

Fig. 10.3   The occipital bone (from below).

2.   The *petromastoid part* forms the posterior portion of the bone and can be divided into two parts:

(a)   The *mastoid portion* which continues into a conical projection called the *mastoid process*, containing air cells.

(b)   The *petrous portion*, between the occipital bone and the sphenoid, which contains the structures forming the internal ear.

3.   The *tympanic part* is a curved plate lying below the squamous

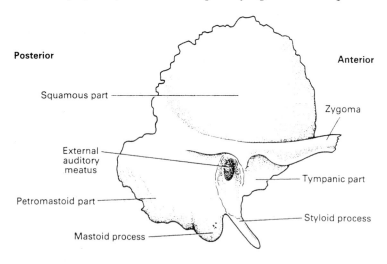

Fig. 10.4   The temporal bone.

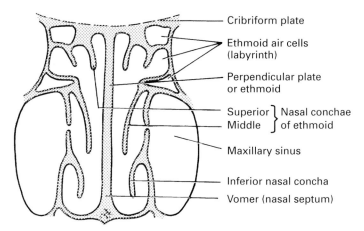

Cribriform plate

Ethmoid air cells
(labyrinth)

Perpendicular plate
or ethmoid

Superior ⎤ Nasal conchae
Middle ⎦ of ethmoid

Maxillary sinus

Inferior nasal concha

Vomer (nasal septum)

Fig. 10.5 Diagram of a section of the ethmoid bone (shown in colour).

part and in front of the mastoid process. It contains the external auditory meatus.

4. The *styloid process* projects downwards and forwards from the underneath of the bone.

The *ethmoid bone* (Fig. 10.5) is very light and irregular in shape and consists of three parts:

1. The *cribriform plate*, a small horizontal plate which is perforated with many fine openings, or foramina, for the passage of the *olfactory nerves*, which transmit the sense of smell.
2. The *perpendicular plate*, which descends from the cribriform plate, and forms the upper part of the nasal septum, which divides the nasal cavity into two.
3. Two *labyrinths*, each consisting of a number of thin-walled *ethmoidal air cells* which communicate with the nasal cavity and may become infected from it. Two thin plates of bone called the *superior and middle nasal conchae*, jut out into the nasal cavities from the spongy labyrinths.

The *sphenoid bone* (Fig. 10.6) is situated at the base of the skull, in front of the temporal bones. It is shaped rather like a bat with outstretched wings. The body contains two large air sinuses, which communicate with the nasal cavity, and a deep depression, known as the *hypophyseal fossa*, which contains the *hypophysis cerebri* or pituitary gland. The *greater and lesser wings* are perforated by many openings for the passage of nerves and blood vessels.

The *inferior nasal conchae* are curved plates of bone which lie in

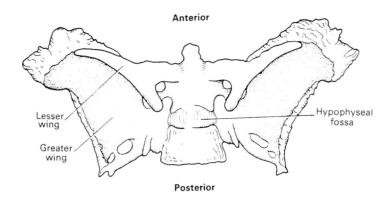

Fig. 10.6   The sphenoid bone.

the walls of the nasal cavity, below the superior and middle nasal conchae of the ethmoid bone (see Fig. 10.5).

The *lacrimal bones* are the smallest and most fragile of the cranial bones and form part of the walls of the orbits. Each is grooved to contain the *lacrimal sac* and *nasolacrimal duct*, by which the *lacrimal fluid*, or tears, which washes constantly over the surface of the eye, is carried into the nasal cavity.

The *nasal bones* are two small oblong bones which together form the bridge of the nose.

The *vomer* is a flat bone which forms the lower part of the septum of the nose.

The Bones of the Face

The bones of the face are listed in Box 10.3.

The bones of the face

- The maxillae
- The mandible
- Two zygomatic bones
- Two palatine bones
- The hyoid bone

Box 10.3

The *maxillae* are the largest bones of the face (excepting the mandible) and by their union in the midline form the whole of

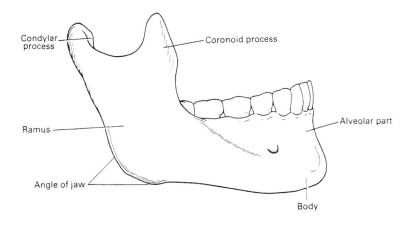

Fig. 10.7 The mandible.

the upper jaw. They carry the upper teeth embedded in a ridge of bone called the *alveolar process*. The *palatine process* of the maxilla is a horizontal projection which forms a considerable part of the floor of the nasal cavity and the roof of the mouth. The *maxillary sinus* is an air-filled cavity within the body of the bone which communicates with the nasal cavity and which may become infected following nasal infection.

The *mandible* (Fig. 10.7) is an irregular bone and is the only movable bone in the head. It forms the lower jaw and carries the lower teeth embedded in the *alveolar part*. The *rami* are vertical projections which carry a *condylar process* which articulates with the temporal bone and a *coronoid process* which provides attachment for muscle. The upright and horizontal portions meet to form the angles of the jaw.

The *zygomatic bones* are irregular bones which form the prominence of the cheek and part of the walls of the orbit. The *temporal process* articulates with the zygomatic process of the temporal bone to form the *zygomatic arch*.

The *palatine bones* are irregular bones which form part of the hard palate, the lateral wall of the nasal cavity and the floor of the orbit.

The *hyoid bone* (Fig. 10.8) is U-shaped and lies at the base of the tongue, to which it gives attachment. It does not articulate with any other bone but is connected by ligaments to the styloid processes of the temporal bone (see Fig. 10.4).

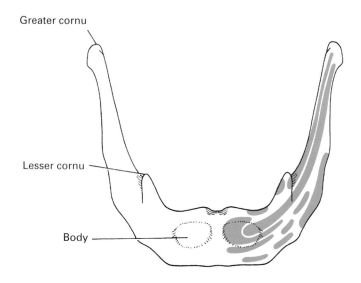

Greater cornu

Lesser cornu

Body

Fig. 10.8   The hyoid bone.

THE BONES OF THE TRUNK

The bones of the trunk are listed in Box 10.4.

The bones of the trunk

- The sternum
- The ribs
- The vertebral column

Box 10.4

The Sternum

The sternum (Fig. 10.9) is a long flat bone which runs down the front of the thorax close under the skin. Its upper end supports the *clavicles* and it also articulates with the first seven pairs of ribs. The bone is in three parts:

1.   The *manubrium* is triangular in shape and its lower border is covered with a thin layer of cartilage for articulation with the upper end of the body.
2.   The *body* is longer and narrower than the manubrium. Where

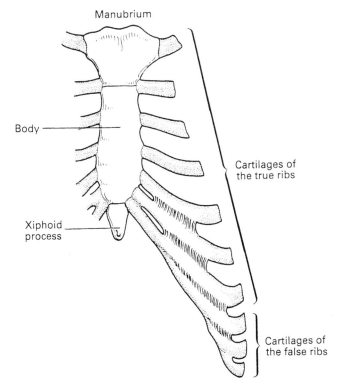

Fig 10.9 The sternum and the costal cartilages

it joins the manubrium there is a small notch which accommodates the cartilage of the second rib.

3. The *xiphoid process* is small and variable in shape and may
not become completely ossified.

The Ribs

The ribs are arched bones which are connected behind with the
vertebral column. There are usually twelve pairs, the first seven of
which are attached to the sternum by the *costal cartilages* and are
known as the *true ribs*. The remaining five are called *false ribs*. Of
these the upper three are joined to the cartilage of the rib immediately above and the lower two are free at their anterior ends and
are known as floating ribs.

The ribs form the curved walls of the thorax, sloping downwards

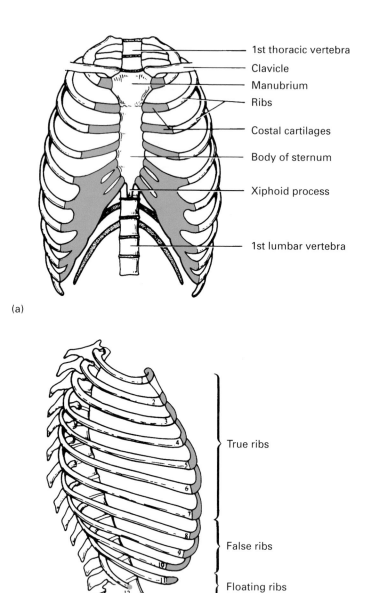

(a)

1st thoracic vertebra

Clavicle

Manubrium

Ribs

Costal cartilages

Body of sternum

Xiphoid process

1st lumbar vertebra

(b)

True ribs

False ribs

Floating ribs

Fig. 10.10   The thoracic cage.

towards the front. They increase in size from above downwards so that the thoracic cavity is roughly cone-shaped (Fig. 10.10).

Each rib is curved and the under surface is grooved for the passage of intercostal arteries, veins and nerves. The *vertebral end* possesses a head, a neck and a tubercle. The head has two smooth facets which articulate with the bodies of the corresponding vertebrae (Fig. 10.11).

The *tubercle* has a small oval facet for articulation with the transverse process of the corresponding vertebra.

The Vertebral Column

The vertebral column consists of a number of irregular bones called the *vertebrae* which are firmly connected to one another but are

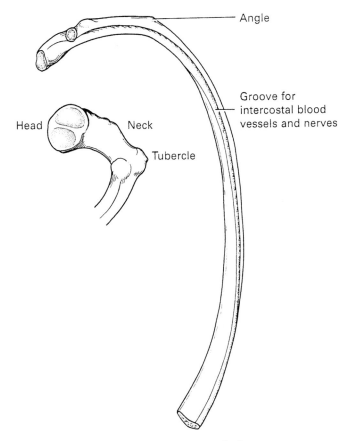

Fig. 10.11 A typical rib.

capable of a limited amount of movement on one another. The column provides a central axis and also protects the spinal cord, which it surrounds. Each vertebra consists of a cylindrical *body* lying to the front, and a *vertebral arch* projecting backwards and enclosing a space called the *vertebral foramen* through which the spinal cord passes. The arch has a *spinous process* directed backwards and downwards and two *transverse processes* which project laterally. These are for the attachment of muscles and ligaments. On the undersurface of the arch is a notch for the passage of spinal nerves and vessels. Each has four articular processes, two above and two below, which meet the corresponding processes of adjoining vertebrae.

The wide parts of the arch carrying the spinous process are known as the *laminae*; this is the part removed in the operation of laminectomy to relieve pressure on the spinal cord caused by disease or following injury.

The bodies of adjoining vertebrae are firmly connected to one another by discs of fibrocartilage called *intervertebral discs*. Each disc has an outer ring of fibrocartilage and an inner core called the *nucleus pulposus*. It is possible for the outer ring to become weakened and for the nucleus pulposus to irritate an adjoining nerve root causing pain.

The vertebrae are divided into five groups (Box 10.5).

The vertebrae

- Seven cervical vertebrae
- Twelve thoracic vertebrae
- Five lumbar vertebrae
- Five sacral vertebrae
- Four coccygeal vertebrae

Box 10.5

The seven *cervical vertebrae* are the smallest and can be easily identified because their transverse processes are perforated by foramina for the passage of the vertebral arteries (Fig. 10.12). The spinous process is forked and gives attachment to muscles and ligaments.

The first cervical vertebra is called the *atlas*. It has no body or spine but consists of a ring of bone with two facets which articulate

Fig. 10.12   A typical cervical vertebra.

with the occipital bone (Fig. 10.13). A ligament called the transverse ligament divides the ring of the atlas into two.

The *axis*, or second cervical vertebra, carries a strong tooth-like process called the *dens* (or odontoid process) which projects

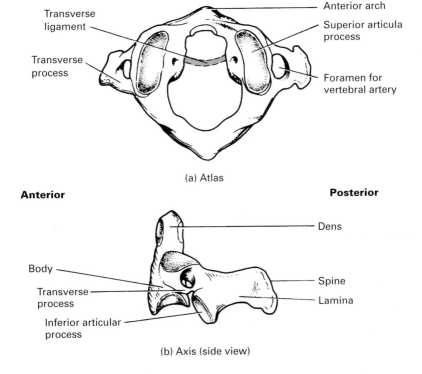

Fig 10.13   The first and second cervical vertebrae.

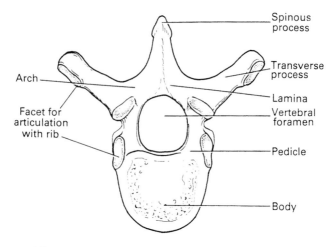

Fig. 10.14   A typical thoracic vertebra (from above).

upwards from the body and provides a pivot on which the atlas, and therefore the skull, rotates, allowing a turning movement of the head (Fig. 10.13). The dens is retained in position by the transverse ligament of the atlas, behind which lies the spinal cord.

The seventh cervical vertebra is distinguished by a very long spinous process which can be seen and felt through the skin at the base of the neck and which is not forked.

The twelve *thoracic vertebrae* are larger than the cervical vertebrae and they increase in size from above downwards. The *body* is roughly heart-shaped and has two distinguishing features:

Fig. 10.15   A typical thoracic vertebra (side view).

Fig. 10.16 A lumbar vertebra.

1.   Additional facets on each side to articulate with the head and tubercle of the corresponding rib (Fig. 10.14).
2.   Long pointed spinous processes which project downwards and backwards (Fig. 10.15).

The heads of the ribs lie between the vertebrae and articulate with one facet on the vertebra above and one on the vertebra below.

The five *lumbar vertebrae* are the largest vertebrae and have no facets for articulation with the ribs. The spinous processes are large and strong and give attachment to the muscles (Fig. 10.16).

The five *sacral vertebrae* are fused together to form a large bone,

Fig. 10.17 The sacrum.

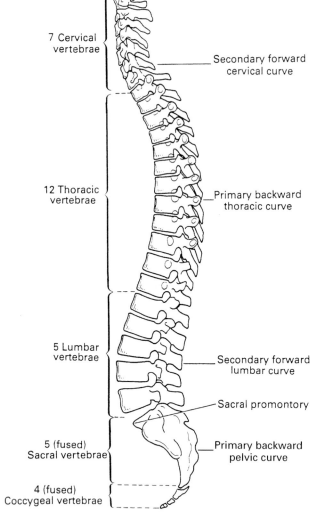

7 Cervical
vertebrae

Secondary forward
cervical curve

12 Thoracic
vertebrae

Primary backward
thoracic curve

5 Lumbar
vertebrae

Secondary forward
lumbar curve

Sacral promontory

5 (fused)
Sacral vertebrae

Primary backward
pelvic curve

4 (fused)
Coccygeal vertebrae

Fig 10.18   The curves of the spine.

the sacrum (Fig. 10.17), which is triangular in shape and forms a wedge between the two hip bones with which it articulates. The pelvic surface of the bone is concave and the anterior projection at the upper end is known as the *sacral promontory* (see Fig. 10.18). The vertebral foramen found in the other vertebrae is here called the *sacral canal*, and from it four openings allow for the passage of nerve roots.

The *coccyx* is a small triangular bone which consists of four vertebrae fused together. It articulates with the sacrum and the joint allows slight movement backwards and forwards which increases the size of the pelvic outlet during childbirth.

The vertebral column is the main support of the head and trunk as well as giving protection to the spinal cord. When viewed from the side it has four curves (Fig. 10.18). The thoracic and pelvic curves are termed primary curves as they are present during fetal life. The cervical and lumbar curves are secondary as they appear or are accentuated when the child begins to hold up its head and sit up (cervical) and when it begins to stand and walk (lumbar).

There is only limited movement between any two adjoining vertebrae but there is considerable movement in the vertebral column as a whole. The intervertebral discs cushion any jarring which may occur; as for example jumping from a height and landing on the feet. The curves of the spine enable it to bend without breaking, but a blow on the column is more likely to cause a fracture or a dislocation because the vertebrae are so firmly united to one another.

Chapter 11
Bones of the Limbs

The bones of the upper limb are listed in Box 11.1.

The bones of the upper limb

- The scapula
- The clavicle   } forming the shoulder girdle
- The humerus
- The radius
- The ulna   } forming the forearm
- Eight carpal bones
- Five metacarpal bones
- Fourteen phalanges

Box 11.1

The *scapula* (Fig. 11.1) is a triangular flat bone; it lies over the ribs at the back of the thorax, but does not articulate with them. It is held in place by muscles which attach it to the ribs and vertebral column. This arrangement gives great freedom of movement to the shoulder girdle, making it possible to reach widely both forwards and backwards and to either side of the body. It is not often broken by falls as it is embedded in muscle. It has three borders and three angles; the lowest is spoken of as the angle because it is the sharpest and is easily felt. The front surface is concave to fit over the ribs; the posterior surface is convex and carries a projecting ridge of bone known as the *spine of the scapula*, which gives attachment to muscles and forms two depressions or fossae, one above and one below it.

The outer angle carries a shallow socket known as the *glenoid cavity*, which receives the head of the humerus to form the shoulder joint. Above this two processes project:

1. The *acromion*, the larger of the two, which overlaps the socket and articulates with the clavicle to form the shoulder girdle.
2. The *coracoid process*, which juts forward and is like a hook.

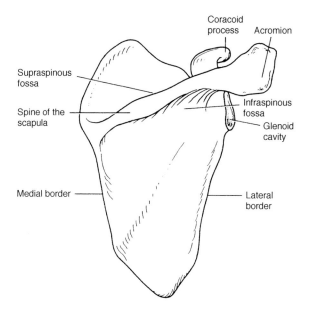

Fig. 11.1 The scapula.

Both of these can easily be felt. They serve to give attachment to muscles and also help to keep the head of the humerus in place, preventing upward dislocation.

The *clavicle* (Fig. 11.2) is a long bone, roughly S-shaped. It articulates with the sternum at its inner or sternal extremity and with the scapula at its outer or acromial extremity. The two extremities are easily distinguishable from one another. The inner extremity is roughly like a pyramid in shape, while the outer is flatter and very similar in shape and form to the acromion process of the scapula, with which it articulates. The bone lies close under the skin and is easily felt along its whole course; starting from the sternal extremity it curves first forwards and then backwards. It keeps the scapula in position and when it is broken the shoulder drops forwards and downwards. It is the only bony link between

Fig. 11.2 The clavicle.

the bones of the upper limb and the axial skeleton, as the scapula does not articulate with either the ribs or vertebral column. It is a bone which is not found in the skeleton of many four-footed animals, as it only becomes necessary to fix the scapula when the limb is moved outwards from the trunk. The bone is easily broken by falls on the shoulder, as it is compressed between the sternum and the point of impact; it is, in fact, better that it should break than that there should be an injury at the root of the neck, where there are many important structures, or about the actual shoulder joint, where injury would be likely to limit subsequent movement.

The *humerus* (Fig. 11.3) is the largest and longest bone of the upper limb. The upper extremity has a hemispherical *head*, covered with hyaline cartilage, which articulates with the glenoid cavity of the scapula, forming the shoulder joint. The *anatomical neck* forms a slight constriction adjoining the head and the *greater* and *lesser tubercles* lie below the neck and give attachment to muscles. Between the tubercles a deep groove accommodates one of the

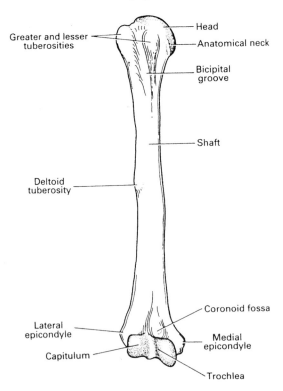

Fig. 11.3   The humerus (front view).

tendons of the biceps muscle. The shaft of the humerus has many roughened surfaces for the attachment of muscles, the most marked being the *deltoid tuberosity* on the outer side, which gives insertion to the deltoid muscle. A groove running obliquely round the shaft carries the radial nerve, one of the three main nerves of the upper limb.

The lower end of the humerus is divided into articular and non-articular portions. The *articular portion*, together with the radius and the ulna, forms the elbow joint. It is divided by a shallow groove into the *capitulum*, a rounded projection which articulates with the radius, and the *trochlea* which is shaped rather like a pulley and which articulates with the ulna. The *non-articular portion* has two *epicondyles* which give attachment to muscles. There are also two deep hollows; the posterior one is called the *olecranon fossa* because it accommodates the olecranon process of the ulna when the elbow is extended and the *coronoid fossa*, which is on the anterior surface, provides for the coronoid process of the ulna when the elbow is flexed.

The *radius* (Fig. 11.4) is the outer bone of the forearm. The upper end is smaller and has a disc-shaped *head* with a hollowed upper surface to articulate with the capitulum of the humerus. The head also articulates with the ulna. The *neck* of the radius is the constricted portion below the head and on the ulna side there is a projection called the *radial tuberosity* which gives insertion to the biceps muscle. The *shaft* of the bone has a sharp ridge facing the ulna and from it a sheet of fibrous tissue called the *interosseous membrane* runs to the ulna, connecting the two bones. The lower end of the radius is the wider part and takes part in the formation of the wrist joint; it also has a projection called the *styloid process* which can be felt at the base of the thumb.

The *ulna* (Fig. 11.4) lies on the inner side of the forearm. The upper end is shaped like a hook and has two large projections; the *olecranon* fits into the olecranon fossa of the humerus when the arm is straight and its upper border forms the point of the elbow. It provides insertion for the tendon of the triceps muscle. The *coronoid process* is smaller and projects forwards. These two processes help in the formation of the *trochlear notch* which articulates with the trochlea of the humerus. The *radial notch* is a depression on the upper part of the coronoid process which articulates with the head of the radius and allows the turning movement of the hand. When rotation of the hand takes place the lower end of the radius is carried round the lower end of the ulna so that the shafts of the two bones cross each other in the middle of the forearm.

Fig. 11.4   The radius and the ulna.

The *shaft* of the ulna, like that of the radius, carries a sharp ridge for the attachment of the interosseous membrane which lies between the two bones.

The lower end of the ulna has a rounded part, known as the *head*, and a projection called the *styloid process*. The head articulates with the ulnar notch of the radius. The styloid process gives attachment to a ligament of the wrist joint; it may be felt below the skin of the wrist and is sometimes quite prominent.

The *carpal bones* (or *carpus*) (Fig. 11.5) comprise eight bones which are arranged in two rows of four. The bones of the upper row are named the *scaphoid*, the *lunate*, the *triquetral* and the *pisiform*. The first three articulate with the radius. The bones of the lower row are called the *trapezium*, the *trapezoid*, the *capitate* and the *hamate*. The surface of the carpus which forms the palm has a deep cavity, the carpal groove, which has a fibrous band across it and which accommodates the median nerve and some of the tendons to the hand. This is known as the *carpal tunnel*.

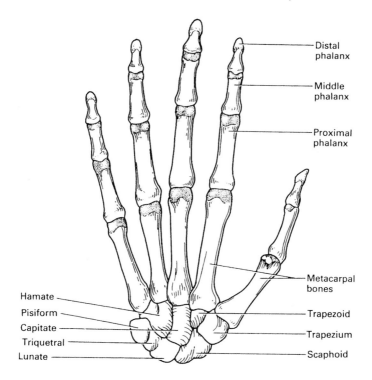

Distal
phalanx

Middle
phalanx

Proximal
phalanx

Metacarpal
bones

Trapezoid

Trapezium

Scaphoid

Hamate
Pisiform
Capitate
Triquetral
Lunate

Fig 11.5   The hand.

The *metacarpal bones* (or metacarpus) are five miniature long bones running across the palm of the hand. The *bases* of the bones articulate with the lower bones of the carpus; the heads articulate with the phalanges. The first metacarpal, which articulates with the two phalanages which form the thumb, can be moved more freely than the other four and can be opposed to each finger in turn, thereby increasing the power of the grip.

The *phalanges* are also miniature long bones, three for each finger and two for the thumb.

BONES OF THE LOWER LIMB

The bones of the lower limb are listed in Box 11.2.

The *hip bone* (Fig. 11.6) is a large, irregularly shaped bone which articulates in front with the corresponding bone of the opposite side. Each bone consists of three parts, the *ilium*, the *ischium* and the *pubis* which are united at the deep cavity on the outer aspect of the

---

The bones of the lower limb

- The hip bone, which forms part of the pelvis
- The femur
- The patella
- The tibia } forming the leg
- The fibula
- Seven tarsal bones
- Five metatarsal bones
- Fourteen phalanges

Box 11.2

---

bone called the *acetabulum*. Full ossification is not completed until between the ages of fifteen and twenty-five and before the bones are united by cartilage.

The *ilium* includes the upper part of the acetabulum and the expanded flattened area of bone above it which is called the *iliac crest* and which gives attachment to the lateral muscles of the abdominal wall. The crest projects a little beyond the lower part

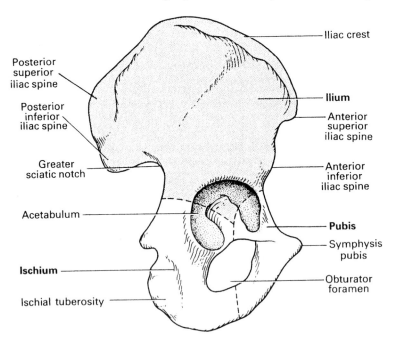

Fig. 11.6   The hip bone.

of the bone and ends in front at the *anterior superior iliac spine* which is easily felt at the lateral end of the fold of the groin. The *posterior superior iliac spine* underlies a small dimple which is readily seen on the lower part of the back. The ilium also carries the *anterior and posterior inferior iliac spines* which provide attachment for powerful muscles. At the back there is an articular surface for articulation with the sacrum and below this there is a notch called the *greater sciatic notch* for the passage of part of the sciatic nerve.

The *ischium* forms the lower posterior part of the hip bone. It carries the *ischial tuberosity* which gives attachment to muscles and supports the body in the sitting position.

The *pubis* forms the anterior part of the hip bone and meets the pubis of the opposite side in a cartilaginous joint called the *symphysis pubis*. The pubis consists of a *body* which enters into the symphysis and two branches, one running up to join the ilium and the other down to join the ischium. Between these branches and the ischium is a large opening called the *obturator foramen* which is filled with a sheet of fibrous tissue.

The *acetabulum* is the deep socket in the centre of the lower part of the hip bone into which the head of the femur fits.

The *pelvis* is a bony ring composed of the two hip bones and the sacrum and coccyx behind. It is divided into the greater (false) and the lesser (true) pelvis by the *linea terminalis* and the promontory of the sacrum. The *greater pelvis* is the upper expanded portion bounded on each side by the ilium and at the back by the base of the sacrum. The *lesser pelvis* consists of a short curved canal deeper at the back than the front. The female pelvis is shorter than the male and is also wider. The brim of the female pelvis is larger and

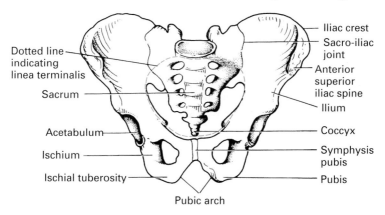

Fig. 11.7 The male pelvis.

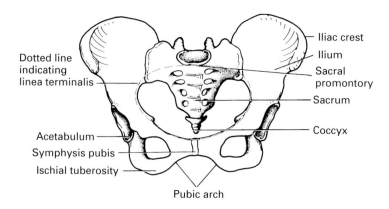

Fig. 11.8 The female pelvis.

more nearly circular than that of the male which is more typically heartshaped (Figs. 11.7 and 11.8).

The *femur* (Fig. 11.9) is the longest and strongest bone in the body. The upper end of the bone has a hemispherical *head* which articulates with the acetabulum of the hip; near its centre is a small depression called the *fovea* (see Fig. 12.6), which gives attachment to the ligament of the head of the femur; this ligament runs to the base of the acetabulum. The *neck of the femur* projects at a marked angle from the shaft and enables free movement of the hip joint. Where the neck joins the shaft there are two processes, the *greater* and *lesser trochanters* which give attachment to muscles. The greater trochanter is on the outer side and can be felt under the skin. The *shaft* of the femur is thinnest in the middle and widens considerably at the lower end. The posterior border is formed by a roughened ridge called the *linea aspera* which gives attachment to muscles.

The lower end of the femur is widely expanded so that there is a good area for the transmission of body weight to the tibia. It has two prominent condyles which articulate with the tibia; they are separated at the back by a deep gap called the *intercondylar fossa* and united in front by a smooth surface which articulates with the patella. Above the condyles at the back of the bone is the *popliteal surface* which forms part of the *popliteal fossa*, containing blood vessels and nerves.

The *patella* is situated in front of the knee joint in the tendon of the quadriceps muscle, which straightens the knee. Bones which develop in tendons in this way are called *sesamoid* bones. The patella is flattened and triangular in shape with the apex downwards. The posterior surface is smooth and articulates with the

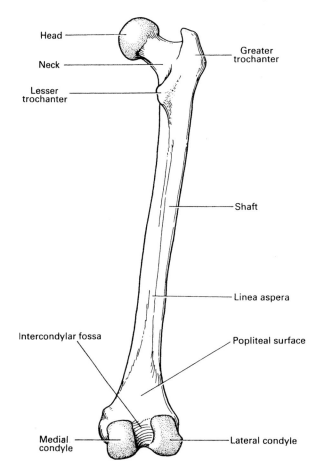

Fig. 11.9 The femur (back view).

condyles of the femur; the anterior surface is roughened and is separated from the skin by a sac, similar to synovial membrane, called a *bursa*.

The *tibia* (Fig. 11.10) is the stronger of the two bones of the leg and lies on the inner, or medial, side. The upper end is greatly expanded to provide a good bearing surface of the body weight. It has two prominent masses, called the *medial* and *lateral* condyles which are smooth and articulate with the condyles of the femur. Between the condyles is a roughened area which gives attachment to the ligaments and cartilages of the knee joint. Below the condyles is a smaller projection called the *tuberosity of the tibia* which gives

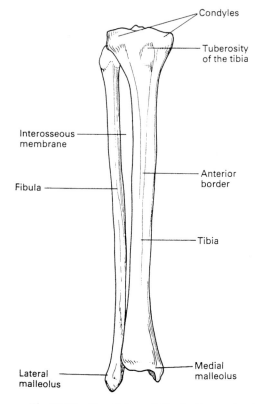

Fig 11.10   The tibia and fibula (front view).

attachment to the *ligamentum patellae*. The lateral condyle has a small circular facet for articulation with the upper end of the fibula.

The *shaft* of the tibia is roughly triangular in cross-section. The *anterior border* lies immediately under the skin and can be felt as the shin. A second border faces the fibula and gives attachment to the *interosseous membrane* which connects tibia and fibula, as the ulna and radius are connected in the forearm.

The *lower end* of the bone is slightly expanded and projects downwards to form the *medial malleolus,* on the inner side of the ankle, which articulates with the talus. The bone also articulates with the fibula.

The *fibula* (Fig. 11.10) is very slender compared with the tibia and lies on the outer side of the leg. The *head* of the fibula has a circular facet which articulates with the lateral condyle of the tibia but does not enter into the formation of the knee joint. The *shaft* is slender and ridged and one of the ridges gives attachment to the inteross-

Phalanges

Metatarsal
bone

Cuboid

Cuneiform bones { Medial
Intermediate
Lateral

Navicular

Talus

Calcaneus

The foot.

Fig. 11.11   The foot.

eous membrane connecting the tibia and fibula. The *lower end* of the
fibula projects downwards to a lower level than the tibia and
carries the bony prominence on the outside of the ankle known
as the *lateral malleolus* which articulates with the talus.

The *tarsal bones* (or *tarsus*) (Fig. 11.11) comprise seven bones
which make up the posterior half of the foot. The *talus* is the
principal connecting link between the foot and the leg and forms
an important part of the ankle joint. The *calcaneus* is the largest and
strongest of the tarsal bones; it projects backwards to form the
prominence of the heel and to provide a lever for the muscles of
the calf which are inserted into its posterior surface. The *navicular
bone* lies between the talus and the three cuneiform bones. The
three *cuneiform bones* are wedge-shaped and articulate with the
navicular and the first three metatarsal bones. The *cuboid bone* is

situated between the calcaneus and the fourth and fifth metatarsal bones.

The *metatarsal bones* (or metatarsus) are five miniature long bones, resembling the metacarpus. The *bases* articulate with the cuneiform and the cuboid bones. The *heads* articulate with the phalanges. As in the hand the *phalanges* of the foot correspond in number and arrangement, there being only two in the big toe and three in each of the other toes.

The Arches of the Foot

The foot has two main functions: to support the weight of the body and to propel the body forward when walking. To fulfil these functions the foot has two longitudinal arches, the *medial arch* which is particularly resilient and the *lateral arch* which is strong and allows only limited movement. There is also a series of *transverse arches*.

The arches of the foot give spring to the walk. They are helped in this by the strong ligaments, tendons and muscles which cross the sole of the foot and which, if stretched, may allow lowering of the medial longitudinal arch and eventually changes in the bones themselves, a condition known as 'flat foot'.

Chapter 12
Joints or Articulations

A joint or articulation is formed wherever two bones meet, but not all junctions of bones allow movement. This chapter looks at the different types of joint in the body and the types of movement which are possible at each. There are three groups of joints (Box 12.1).

Groups of joints

- **Fibrous** or fixed joints
- **Cartilaginous** or slightly movable joints
- **Synovial** or freely movable joints

Box 12.1

However not all joints fit rigidly into this classification as there are some fibrous joints which are slightly movable (e.g. the joint between the lower ends of the tibia and fibula) and some cartilaginous joints which are barely movable (e.g. the symphysis pubis).

Bones are joined to each other by ligaments, which are usually strong cords of fibrous tissue attached to the periosteum and running from one bone to another. These ligaments are yielding but inelastic and vary in strength and shape according to the work that they have to do. Because they are pliable they allow movement to take place but they are also strong, inelastic and rich in sensory nerves and in this way they protect the joints from excessive movement and strain.

FIBROUS JOINTS

One type of fibrous joint (Fig. 12.1) occurs where the margins of two bones meet and dovetail accurately into one another, separated only by a thin band of fibrous tissue. Joints such as these, which do not normally permit movement, are found between the bones of the cranium and are called *sutures*. In the infant at birth there is a definite line of fibrous tissue between two adjoining bones which allows the edges to glide over one another, enabling the head to be

Fig. 12.1   Diagram of a fibrous joint.

moulded to ease its passage through the birth canal. Other fibrous joints occur where the roots of the teeth articulate with the upper and lower jaws and where there is an interosseous ligament, as in the tibiofibular joint.

CARTILAGINOUS JOINTS

A cartilaginous joint (Fig. 12.2) occurs where the two bony surfaces are covered with hyaline cartilage and connected by a pad of fibrocartilage and by ligaments, which do not form a complete capsule round the joint. A limited degree of movement is possible because the cartilaginous pad can be compressed. The joints between the bodies of the vertebrae and between the manubrium and the body of the sternum are cartilaginous joints.

SYNOVIAL JOINTS

A synovial joint (Fig. 12.3) consists of two or more bones the ends of which are covered with articular hyaline cartilage. There is a joint cavity containing *synovial fluid* which nourishes the avascular articular cartilage and the joint is completely surrounded by a

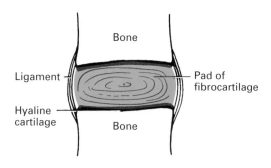

Fig 12.2   Diagrammatic section through a cartilaginous (slightly movable) joint.

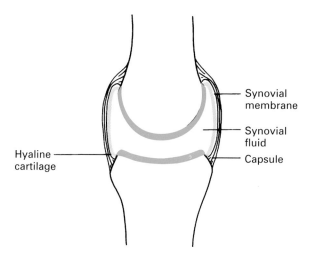

Fig. 12.3 Diagrammatic section through a synovial (freely movable) joint.

fibrous capsule lined with *synovial membrane* which lines the whole of the interior of the joint with the exception of the bone ends, menisci and discs. The bones are also connected by a number of ligaments, and some movement is always possible in a synovial joint even though it may be limited, as in the gliding movement between the adjoining metacarpal bones.

Types of synovial joint

- **Hinge joints**—allow movement in one direction only; the elbow is a good example
- **Pivot joints**—allow rotation; e.g. the radius and ulna at the elbow and the first and second cervical vertebrae allowing turning movements of wrist and head, respectively
- **Condylar joints**—two pairs of articular surfaces allow movement in one direction only, but the articular surfaces may be enclosed in the same or in different capsules; the knee joint is one example
- **Ball-and-socket joints**—these are formed by a hemispherical head fitting into a cup-shaped socket; examples are the hip and shoulder joints
- **Plane joints**—gliding movements are restricted by ligaments or bony prominences; e.g. as in the carpal and tarsal joints

Box 12.2

In some synovial joints the cavity may be divided by an *articular disc* or *meniscus* consisting of fibrocartilage which helps to lubricate the joint, to reduce wear of the articular surfaces and to deepen the joint.

Types of Synovial Joint

The synovial joints are divided into several classes according to the type of movement which occurs (Box 12.2).

JOINT MOVEMENTS

Joint movements are of three kinds; gliding, angular and circular (Box 12.3). These movements are often combined to produce a variety of movements at any one joint.

Movement of joints

- **Gliding movement** occurs, as its name implies, without any angular or rotatory movement.
- **Angular movement** brings about an increase or decrease of the angles between the bones. It includes *flexion* or bending and *extension* or straightening, and also *abduction*, movement away from the midline and *adduction* movement towards the midline.
- **Circular movement** allows *internal rotation*, turning a part on its own axis towards the midline, and *external rotation* away from the midline. *Circumduction* allows movement of a limb through a circle. The words *supination* and *pronation* refer to turning the hand palm uppermost or palm downwards respectively.

Box 12.3

Different types of joint have varying kinds of movement. The ball-and-socket joints are most freely movable allowing all movements described above, excluding supination and pronation. Hinge joints allow flexion and extension only. In gliding joints there is only slight movement increasing the range in all directions, as in the carpal and tarsal bones at wrist and ankle. In the hand the terms adduction and abduction are used to refer to movement to and from the middle line of the part and not from the midline of the body as a whole. Adduction of the thumb brings it to and across the palm of the hand, and adduction of the little finger brings it towards the thumb. In the anatomical position, with the palm facing forward, the little finger is moving away from the midline of the body but towards the midline of the hand.

The joints are movable, but the movements are carried out by the various muscles with which the joints are provided. The muscles, in addition to producing movement, run, as do the ligaments, from bone to bone, and help to hold the bones in position and give support to the joint capsule, as long as their normal tone is sustained. When the muscles are paralysed and limp, the looser joints are dislocated comparatively easily, particularly the freely movable shoulder joint; when the muscles are paralysed, but rigid and shrunken, the joint may become completely immovable. This immobility can be prevented by moving the joint through as wide a range of movement as possible to maintain the elasticity of the muscles. Immobility may also result from the joint surfaces becoming adherent to one another or to the joint capsule as a result of disease or injury affecting the muscle itself.

THE JOINTS OF THE HEAD

The *temporomandibular joint*, between the temporal bone and the head of the mandible, is the only movable joint of the head and it is peculiar in that movement can occur in all three planes: upwards and downwards, backwards and forwards and from side to side.

The sutures between the skull bones have already been described (Chapter 10). However, at the angles of the parietal bones there are unossified membranous areas called *fontanelles* (Fig. 12.4).

The *anterior fontanelle* is the largest and lies at the junction of the two parietal bones with the frontal bone. It is diamond-shaped and does not close completely until the age of fifteen to eighteen months. Dehydration in infancy causes the fontanelle to become depressed which is a serious sign. A large venous sinus runs under the fontanelle, from which it is possible to obtain a specimen of blood and into which intravenous fluids may be given in the infant.

The *posterior fontanelle* lies at the junction of the parietal bones with the occipital bone. It is triangular in shape and closes shortly after birth. Delay in the closure of the fontanelles may be caused by hydrocephalus but also occurs sometimes in the normal child.

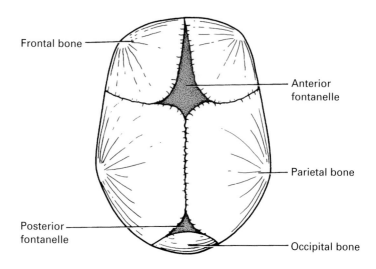

Fig. 12.4   An infant's head at birth, showing the fontanelles.

THE JOINTS OF THE TRUNK

There are joints between all the vertebrae from the second cervical to the sacrum. Cartilaginous joints lie between the vertebral bodies and synovial joints between the vertebral arches. Because the joints are so numerous the spinal column as a whole has considerable movement. *Anterior* and *posterior longitudinal ligaments* pass from the top of the spine to the sacrum giving support. Other ligaments pass between the vertebral arches.

Between the ribs and the vertebrae are the *costovertebral joints* which allow gliding movements; the same type of movement occurs at the *sternocostal joints*.

THE JOINTS OF THE UPPER EXTREMITY

The *sternoclavicular joint* is formed by the sternal end of the clavicle, the manubrium of the sternum and the cartilage of the first rib. It allows a gliding movement of the clavicle.

The *acromioclavicular joint* lies between the acromial extremity of the clavicle and the acromion of the scapula and is usually associated with movements of the shoulder.

The *shoulder joint* is a ball-and-socket joint (Fig. 12.5), and is the most freely movable of the joints in the body. It is formed by the head of the humerus fitting into the small, shallow glenoid cavity.

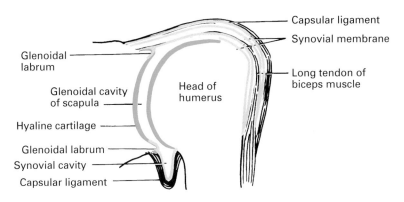

Fig. 12.5 Diagrammatic section through the shoulder joint.

The articular surfaces are covered with articular cartilage and the glenoid cavity is enlarged and deepened by a rim of fibrocartilage, called the *glenoidal labrum*, which runs around it. This lessens the risk of dislocation without limiting the movement as much as a larger and deeper bony socket would do. The bones are held together by a loose capsule of ligaments to give the joint its wide range of movement, but the powerful muscles help to keep the bones in position. The long tendon of the biceps muscle serves as an intracapsular ligament. It runs through the bicipital groove between the tuberosities of the humerus into the joint cavity and, since it arises from the scapula immediately above the glenoid cavity, it tends to hold the articular surfaces in position. Movement of the arm above the level of the shoulder joint is due to movement of the scapula over the back of the thorax.

The *elbow joint* is complicated by the fact that it contains the hinge joint between the humerus and the radius and ulna and the pivot joint between the radius and ulna. There are strong ligaments running between the three bones and a circular ligament called the *annular ligament* which keeps the head of the radius in contact with the radial notch of the ulna. The lower end of the radius also forms a pivot joint with the ulna.

The *wrist joint* is formed by the lower end of the radius and the scaphoid, lunate and triquetral bones. Together with the joints between the carpal bones the movements of flexion, extension, adduction (ulna deviation), abduction (radial deviation) and circumduction can be carried out.

The *metacarpophalangeal joints* are also capable of all the movements described for the wrist joints but the interphalangeal joints are hinge joints, allowing flexion and extension only.

THE JOINTS OF THE LOWER EXTREMITY

The *sacroiliac joint* is a synovial joint which allows a small amount of rotary movement during flexion and extension of the trunk.

The *symphysis pubis* is a cartilaginous joint which moves very little. However during pregnancy the pelvic joints and ligaments are relaxed to allow slightly greater movement.

The *hip joint* (Fig. 12.6) is a ball-and-socket joint formed by the head of the femur fitting into the cup-shaped acetabulum. The joint surfaces are covered with articular cartilage and the acetabulum, like the glenoid cavity, is deepened by a rim of fibrocartilage called the *acetabular labrum*. The *ligament of the head of the femur* is attached to a small, roughened pit called the *fovea* near the centre of the head of the femur, and runs to the acetabulum. The joint has a strong *fibrous capsule* and many ligaments, one of which, the *iliofemoral ligament*, lies across the front of the joint and prevents extension of the hip beyond a straight line with the trunk. A wide range of movement is possible at the hip joint, though when the knee is flexed, flexion of the hip is limited by the contact of the thigh with the anterior abdominal wall; also, when the knee is extended, flexion of the hip is limited by tension of the hamstring muscles.

The *knee joint* (Fig. 12.7) is the largest joint of the body. It is a compound joint: a condylar joint which gives articulation between

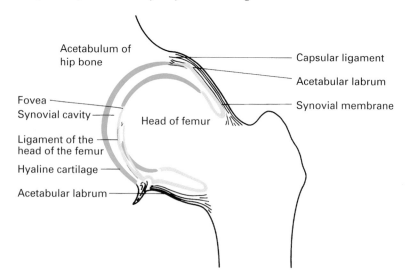

Fig. 12.6 Diagrammatic section through the hip joint.

Fig. 12.7 Diagrammatic section through the knee joint.

the condyles of the femur and the tibia and a plane joint which gives articulation between the patella and the femur. The joint has a *fibrous capsule* into the front of which the patella enters and which is lined with synovial membrane. The *cruciate ligaments* are very strong and cross each other within the joint. They run from the intercondylar area of the tibia to the femur and are partially covered with synovial membrane. The extracapsular ligaments are also strong and thick and help in controlling movement of the joint. The *menisci* (or semilunar cartilages) deepen the surfaces of the upper end of the tibia. They are wedge-shaped, the outer border being thick and convex and the inner thin and concave, and they can be injured as a result of a twisting strain when the knee is flexed. However, if fully exercised they will reform. The movements of the knee joint are mainly flexion and extension, although some rotation can also occur.

The *upper tibiofibular joint* is a synovial plane joint allowing a little gliding movement while at the lower end of the bones there is a small amount of rotation of the fibula during some ankle movements.

The *ankle joint* is a hinge joint formed between the tibia, fibula and talus (Fig. 12.8). The movements are flexion and extension but are usually spoken of as *dorsiflexion*, raising the foot, and *plantar flexion*, raising the heel.

Fig. 12.8   X-rays of the right ankle joint, showing (left) lateral view and (right) anteroposterior view. (1) Tibia; (2) fibia; (3) medial malleolus; (4) lateral malleolus; (5) talus; (6) calcaneum; (7) cuboid.

The joints between the tarsal bones and between the tarsus and metatarsus are gliding joints and movement is limited. The *metatarsophalangeal joints* and the *interphalangeal joints* allow movements similar to the corresponding joints in the hand.

Chapter 13
Structure and Action of Muscle

The muscular system consists of a large number of muscles through which the movements of the body are carried out. Voluntary muscles are attached to bones, cartilages, ligaments, skin or to other muscles by fibrous structures called *tendons* and *aponeuroses*. The individual fibres of voluntary muscle with their sheaths of *sarcolemma* are bound together into bundles by the *endomysium* and are covered by the *perimysium*. The bundles, or *fasciculae*, are bound together by a denser covering called the *epimysium* and these groups form the individual voluntary muscles of the body (see Fig. 4.8). All muscles have a good blood supply from nearby arteries. Arterioles in the perimysium give off capillaries which run in the endomysium and across the fibres. Blood vessels and nerves enter the muscle together at the hilum.

Most muscles have tendons at one or both ends. *Tendons* are made of fibrous tissue and are usually cord-like in appearance, though in some flat or sheet-like muscles the cord is replaced by a thin, strong fibrous sheet called an *aponeurosis*. Fibrous tissue also forms a protective covering or muscle sheath, known as *fascia*.

Where one muscle is attached to another the fibres may interlace, the perimysium of one fusing with the perimysium of the other, or the two muscles may share a common tendon. A third type of connection occurs in the muscles of the abdominal wall where the fibres of the aponeuroses interlace, forming the *linea alba*, which can be seen as a shallow groove above the umbilicus.

When a muscle contracts one end normally remains stationary while the other end is drawn towards it (Figs. 13.1 and 13.2). The end which remains stationary is called the *origin* and that which moves is called the *insertion*. It is not uncommon, however, for a muscle to be used, as it were, the wrong way round so that the insertion remains fixed and the origin moves towards it. The

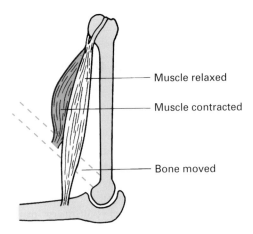

Fig. 13.1 Diagram showing bone movement as muscle contracts.

gluteus maximus provides an illustration. Its origin is in the sacrum and it is inserted into the femur. When the insertion moves towards the origin the flexed thigh is extended; when the body is bent forward at the hips the standing position is regained by movement of the origin towards the insertion. This arrangement economizes on the number of muscles required and further economy is achieved by the placing of muscles so that they can carry out more than one action. Muscles must cross the joint they move; some cross two joints producing movement in both, e.g. the biceps crosses both elbow and shoulder causing flexion of both.

Muscles only act by contracting and pulling; they cannot push, though they can contract without shortening and so hold a joint

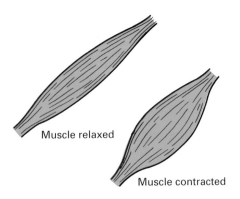

Fig. 13.2 A typical voluntary muscle showing contraction.

firm and fixed in a certain position. When the contraction passes off the muscles become soft but do not lengthen until stretched by the contraction of opposing muscles, known as antagonists.

Muscles never work alone; even the simplest movement requires the action of many muscles. Picking up a pencil requires movement of fingers and thumb, wrist and elbow and possibly of shoulder and trunk as the body leans forward. Each muscle must contract just sufficiently and each antagonist relax equally to allow the movement to take place smoothly without jerking. This concerted action of many muscles is termed *muscle coordination*; any new action involving coordination requires time and practice until the new combination of muscle movement has been acquired and only then can it be carried out without great mental effort and concentration.

The sensory nerve gives 'muscle sense', not a very acute sensation but sufficient to allow awareness of contraction and relaxation in muscle. There is no awareness of this sensation until a conscious effort is made to relax or contract a muscle at which time the previous degree of concentration becomes obvious. Under normal circumstances the muscles are in a state of partial contraction known as *muscle tone*; it is because of muscle tone that a position can be maintained for long periods without exhaustion. This is dependent on a mechanism whereby different groups of muscle fibres contract and relax in turn giving periods of rest and activity to each group. The muscles having the highest degree of tonicity in man are those of the neck and back.

Contraction of Muscle

Composition of muscle

- 75% Water
- 20% Protein
- 5% Mineral salts, glycogen and fat

Box 13.1

The composition of muscle is shown in Box 13.1. Muscle contraction occurs as a result of nerve impulses. The nerve impulses, which are electrical, are transmitted to the muscle cells by chemical means and this is accomplished by the *neuromuscular junction*

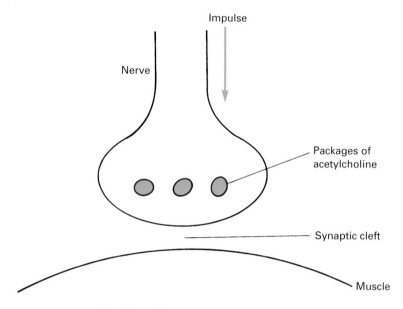

Fig. 13.3   The neuromuscular junction.

(Fig. 13.3). Nerve impulses arrive at the neuromuscular junction, which contains small packages of acetylcholine. Acetylcholine is released into the space between the nerve and the muscle, known as the *synaptic cleft*, and when the acetylcholine attaches to the muscle cell it causes depolarization (see Chapter 5) and electrical activity spreads over the muscle cell leading to contraction. In order for muscle fibres to contract energy is required and this is obtained from the oxidation of food, particularly carbohydrates. During digestion carbohydrates are broken down to a simple sugar called glucose. The glucose which is not required immediately by the body is converted to glycogen and is stored in the liver and in the muscles. Muscle glycogen provides the source of heat and energy for muscular activity (Fig. 13.4). During the oxidation of glycogen to carbon dioxide and water a compound is formed which is rich in energy. This compound is called adenosine triphosphate (ATP). When it is necessary for muscular contraction, the energy from ATP can be released as the compound changes to adenosine diphosphate (ADP). During the oxidation of glycogen pyruvic acid is formed. If plenty of oxygen is available, as it usually is during ordinary movement, pyruvic acid is broken down to carbon dioxide and water, and during the process energy is released, which is used to make more ATP. If insufficient oxygen

Fig. 13.4 Changes during muscle contraction.

is available the pyruvic acid is converted to lactic acid, which accumulates and produces muscle fatigue.

Skeletal muscle is also known as *striated (striped) muscle* (see Chapter 4) as a result of its appearance when viewed under a microscope. The striations are due to the structure of the proteins of which the muscle is composed. These proteins are called *actin* and *myosin*. When muscle is contracted the striations are narrower and this is thought to result from the movement of one protein

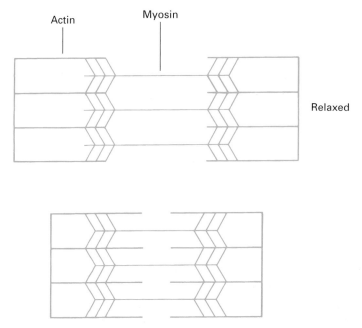

Fig. 13.5 The sliding filament theory.

relative to the other in what is described as the *sliding filament theory* (Fig. 13.5). One of the proteins (myosin) has projections which, with the expenditure of energy, allow it to 'walk' along the other protein, thus leading to contraction when the muscle is stimulated by an electrical impulse.

During violent exercise more oxygen is brought to the muscles but, even so, not enough oxygen reaches the muscle cells, particularly at the beginning of the effort. Lactic acid accumulates and diffuses into the tissue fluid and the blood. The presence of lactic acid in the blood stimulates the respiratory centre and the rate and depth of respiration are increased. This continues, even after the exercise is over, until sufficient oxygen has been taken in to allow the cells of the muscles and the liver to oxidize the lactic acid completely, or convert it to glycogen. This extra oxygen needed to remove the accumulated lactic acid is called the 'oxygen debt' which must be repaid after the exercise is completed.

Chapter 14
The Chief Muscles of the Body

THE MUSCLES OF THE HEAD

The muscles of the head (Fig. 14.1) are divided into two groups according to their functions: (1) the muscles of expression, and (2) the muscles of mastication.

The *muscles of expression* are attached to the skin rather than to the bone, so that they move the skin and change the facial appearance. Circular muscles, called *orbicularis oculi* and *orbicularis oris*, surround the eyes and mouth respectively, closing them. Small muscles raise and lower the eyebrows and upper lids, raise and lower the angles of the mouth and dilate the nostrils, causing a look of surprise, worry, happiness or sorrow. Small muscles also move

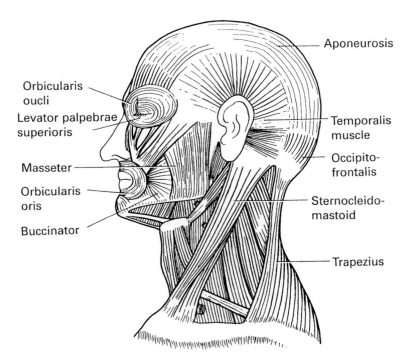

Fig. 14.1 The muscles of the head and neck.

the eyeballs in the orbits (see Chapter 7) both to direct the eyes for sight and also to change the expression.

The *muscles of mastication* move the lower jaw up and down in biting, and also both from side to side and backwards and forwards in chewing. They are the *masseter* running to the angle of the jaw from the zygomatic arch, the *temporalis* muscle, lying over the temporal bone and inserted into the lower jaw, and the smaller muscles which also run from the skull to the lower jaw. These are the muscles which 'lock the jaw' in tetanus.

THE MUSCLES OF THE NECK

The neck contains two large muscles, the sternocleidomastoid and the trapezius (Fig. 14.1).

The *sternocleidomastoid* lies at the side of the neck, running from the sternum and clavicle to the mastoid process and the surface of the temporal bone behind it. When the muscle on one side contracts it draws the head towards the shoulder and turns the head. When both are used together they flex the neck.

The *trapezius* lies over the back of the neck and shoulder and is roughly triangular in shape, with the base joining the spine down the back of the neck and chest, from the occiput, to which it is also attached, downwards; the angle is inserted into the scapula and clavicle, over the top and back of the shoulder. It draws the shoulders back when used as a whole, and also draws the scapula up and down, when the upper and lower portions are used separately (Table 14.1).

THE MUSCLES OF THE TRUNK

The chief muscles of the trunk can be grouped according to their function (Box 14.1).

The muscles of the trunk

- Muscles moving the shoulder
- Muscles of respiration
- Muscles forming the abdominal wall
- Muscles moving the hip
- Muscles moving the spine
- Muscles of the pelvic floor

Box 14.1

Table 14.1 Muscles of the Neck

Name	Position	Origin	Insertion	Action
Sternocleidomastoid	The side of the neck	Sternum and clavicle	Mastoid process	Used separately, turn the head to the side and tilt to same shoulder; used together, flex the neck
Trapezius	The back of the neck and shoulder	The occiput and the spines of the thoracic vertebrae	The clavicle and the spine of the scapula	Draws the scapula back, bracing the shoulders; raising and lowering the shoulder and extending the neck by exerting pull on the occiput

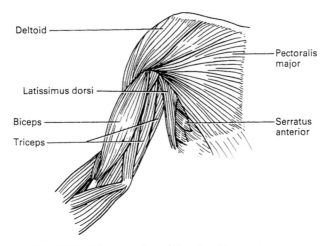

Fig. 14.2   The muscles of the shoulder and arm.

Muscles Moving the Shoulder

The chief muscles moving the shoulder are the powerful muscles covering the back and front of the chest. They include the pectoralis major, the trapezius (see p. 170), the latissimus dorsi and the serratus anterior (Fig. 14.2). The *pectoralis* covers the front of the chest, running from the sternum out to the humerus. The *latissimus dorsi* covers the back of the chest and abdomen, running from the thoracic and lumbar vertebrae and iliac crest out to the humerus (Fig. 14.3). These muscles form the muscle in front of and behind the armpit. The *serratus anterior* runs round the side wall of the thorax from the ribs in front to the vertebral edges of the scapula under which it passes (Table 14.2).

Muscles of Respiration

The chief muscles of respiration are listed in Box 14.2.

The muscles of respiration

- The diaphragm
- The external intercostals
- The internal intercostals

Box 14.2

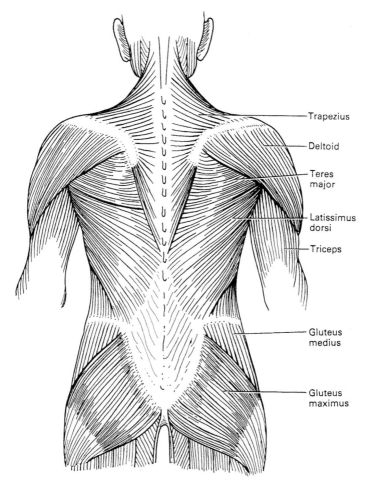

Fig. 14.3 The muscles of the back.

The *diaphragm* is a dome-shaped sheet of muscle dividing the thorax from the abdomen. The border is of muscle, while the centre is a sheet of fibrous tissue or aponeurosis. The muscle arises from the tip of the sternum, the lower ribs and their cartilages, and the first three lumbar vertebrae, and is inserted into the central aponeurosis. Three openings in the diaphragm allow for the passage of the oesophagus, aorta and inferior vena cava, together with lesser structures such as the vagus nerve and thoracic duct, which accompany the oesophagus and aorta respectively. When the muscle fibres contract the dome of the diaphragm is flattened

Table 14.2 Muscles Connecting the Upper Limb with the Trunk

Name	Position	Origin	Insertion	Action
Pectoralis major	Front of chest	Sternum, clavicle and cartilages of true ribs	Humerus (bicipital groove)	Adducts the shoulder, draws the arm across the front of the chest. Internal rotation of shoulder
Latissimus dorsi	Crosses the back from the lumbar region to the shoulder	Lower thoracic, lumbar and sacral vertebrae and iliac crest	Humerus (bicipital groove)	Adducts the shoulder, draws the arm backwards and downwards, as in bell-pulling and rowing, and internal rotation of shoulder
Serratus anterior	Over the sides of the thorax and under the scapula at the back	The front of the eight upper ribs	The medial border of the scapula	Draws the scapula forwards; antagonistic to the trapezius

Fig 14.4 The diaphragm.

and lowered, which increases the depth of the thorax from top to bottom (Fig. 14.4).

The *external intercostal* muscles lie between the ribs, their fibres running downwards and forwards from one rib to the rib below. Each originates from the lower border of the rib above and is inserted into the upper border of the rib below. Their action is to draw the ribs upwards and outwards to increase the size of the thorax from side to side and from back to front.

The *internal intercostal* muscles also lie between the ribs under the external intercostals and are antagonistic to them. Their fibres run downwards and backwards from one rib to the rib below. Each originates from the lower border of the rib above and is inserted into the upper border of the rib below. Their action is to draw the ribs downwards and inwards to decrease the size of the thorax from side to side and from back to front, particularly in forced expiration.

Muscles Forming the Abdominal Wall

The chief muscles of the abdominal wall are listed in Box 14.3 and shown in Fig. 14.5.

The *rectus abdominis* forms the anterior abdominal wall, running up from the pubis to the sternum and costal cartilages. Its fibres run straight up and down, hence its name, for rectus means straight. It is divided into two parts by a line of fibrous tissue in the midline of the body called the *linea alba*. It is also crossed by

---

The muscles of the abdominal wall

- The rectus abdominis, forming the front wall
- The external oblique
- The internal oblique ⎫ forming the side wall and lying
- The transversus abdominis ⎬ one under the other
- The quadratus lumborum

Box 14.3

---

lines of fibrous tissue at intervals. These fibrous bands strengthen the muscle and help to prevent stretching.

The *external oblique* muscle forms the outer layer of the side wall. Its fibres run downwards and forwards. It arises from the lower ribs and is inserted into the iliac crest and the inguinal ligament. The *inguinal ligament* forms the firm edge of the abdominal wall across the groin, where the muscles are not inserted into the bone, leaving a gap for muscles, blood vessels and nerves to pass under it into the limb from the trunk. It gives attachment to the muscles

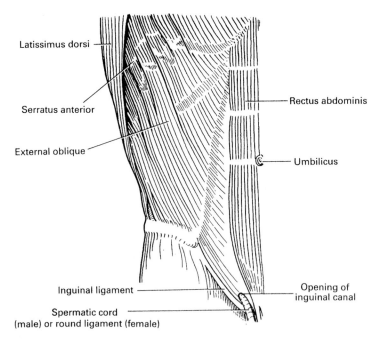

Fig. 14.5 The abdominal wall. Note that the aponeurosis has been cut away to expose the rectus.

here. It is a strong cord of fibrous tissue. Across the front of the abdomen the *external oblique* forms a strong aponeurosis which passes in front of the rectus, joining the linea alba.

The *internal oblique* forms the second layer of the side wall of the abdomen. Its fibres run upwards and forwards. It arises from the iliac crest and the inguinal ligament and is inserted into the lower ribs and their cartilages. It also forms an aponeurosis, passing partly in front and partly behind the rectus, and joining those of the external oblique and transversus abdominis.

The *transversus abdominis* forms the inner layer of the side wall of the abdomen, lying under the internal oblique. Its fibres run straight round the abdominal wall. It arises from the iliac crest and the lumbar fascia, by which it is joined to the lumbar vertebrae. It is inserted into an aponeurosis which runs across the front of the abdomen, behind the rectus and joins the linea alba.

The *quadratus lumborum* forms the posterior wall, running up from the iliac crest to the twelfth rib and upper lumbar vertebrae. It holds the twelfth rib steady during breathing.

The abdominal wall is pierced by a canal in either groin; this is called the *inguinal canal*. It runs obliquely through the muscular coats just above the inguinal ligament near its inner extremity. The canal gives passage to structures: in the male the spermatic cord from the testicle, in the female the round ligaments of the uterus run through it with associated blood vessels and nerves.

Muscles of the Hip

The muscles in the trunk moving the hip are listed in Box 14.4.

The muscles of the hip

- The iliopsoas:
  (a) The psoas
  (b) The iliacus
- The gluteal muscles: maximus, medius, minimus

Box 14.4

The *iliopsoas* muscles cross the front of the groin behind the inguinal ligament (Fig. 14.6). The *psoas* arises from the transverse processes of the lumbar vertebrae, and the *iliacus* from the front surface of the upper part of the ilium. They are both inserted into

Fig. 14.6   The iliopsoas.

the lesser trochanter of the femur. They flex the hip joint and also produce abduction but when the femur is fixed, they bend the trunk forwards.

The *gluteal muscles* form the buttocks (see Fig. 14.3), running from the back of the sacrum and ilium to their insertion in the greater trochanter of the femur and the gluteal ridge below it. They are three in number: the gluteus maximus, medius and minimus. They extend the hip joint and also abduct and laterally rotate the hip (Table 14.3), but when the femur is fixed, they extend the trunk on the lower limb. These muscles are commonly used as the site for intramuscular injections as they are thick and fleshy. Care must be taken to use the upper outer quadrant as the sciatic nerve passes through the other quadrants.

Table 14.3 Trunk Muscles which Move the Hip

Name	Position	Origin	Insertion	Action
Psoas major	Crosses the groin behind the inguinal ligament	Transverse processes of lumbar vertebrae	Lesser trochanter of femur	Flexes the hip
Iliacus	Crosses the groin behind the inguinal ligament with the psoas	Front surface of the iliac bone	Lesser trochanter of femur	Flexes the hip
Gluteal muscles	Cross the back of the hip, forming the buttocks	Posterior surface of ilium and sacrum	Greater trochanter and gluteal line of femur	Extend, abduct and laterally rotate the hip

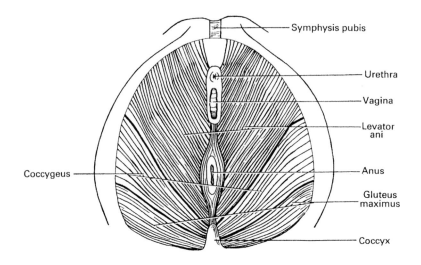

Fig. 14.7   The pelvic diaphragm.

Muscles Moving the Spine

The muscles of the abdominal wall flex and turn the trunk, the rectus flexing and the side muscles turning the thorax on the abdomen. The abdominal muscles also compress the internal organs. The *spinalis muscles* extend the spine. They run up the back of the trunk on either side of the spine. They arise from the back of the iliac crest and the sacrum, and are inserted into the ribs and upper vertebrae.

Muscles of the Pelvic Diaphragm

The pelvic diaphragm consists of muscles which form the support of the pelvic organs (Fig. 14.7); it runs from the pubis in front back to the sacrum and coccyx and out to the ischium on either side. It is like an open book in shape, sloping downwards from back to front and from either side towards the midline. It is composed of the *levator ani* and *coccygeus* muscles. It is pierced in the midline in the female by three openings for the passage of the urethra, the vagina and the rectum. In the male there are only two openings, for the urethra and the rectum respectively.

THE MUSCLES OF THE UPPER LIMB

The muscles of the upper limb may be divided into the muscles of the arm, the forearm and the hand.

The Muscles of the Arm

The muscles of the arm are the largest and strongest of the limb and include those listed in Box 14.5 and shown in Fig. 14.8.

The muscles of the arm

- The biceps
- The triceps
- The deltoid
- The brachialis

Box 14.5

The *biceps* is so called because it has two heads (Latin caput—head). It runs down the front of the arm, where it can easily be felt

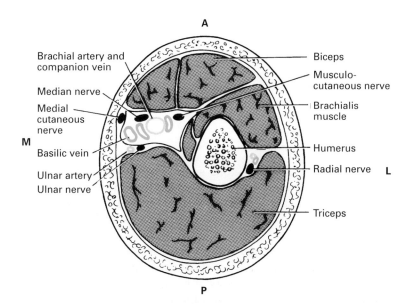

Fig. 14.8 Section of the arm, showing the relation of the muscles to the bone and other structures. L, Lateral; M, median; A, anterior.

when it is contracted. It arises by two heads, one from the glenoid cavity and one from the coracoid process of the scapula, and is inserted into the radial tuberosity in the forearm, crossing the front of the elbow joint. It flexes the elbow and also the shoulder and supinates the hand. Hence the act of supination can be carried out with great force and screws and nuts are made so that a right-handed man tightens them with the action of supination.

The *triceps* is so called because it has three heads. It lies at the back of the arm; it arises by three heads, one from the scapula and two from the humerus, and is inserted into the olecranon of the ulna, at the back of the elbow joint. It extends both elbow and shoulder and is antagonistic to the biceps.

The *deltoid* is triangular in shape. It lies over the shoulder in the position of an epaulette with the base of the triangle forming the origin and attached to the shoulder girdle, just above the shoulder joint. It is inserted into the deltoid tuberosity on the outer side of the humerus. Its action is to abduct the shoulder to a right angle. (To raise the arm above a right angle, the shoulder girdle must move too, and the trapezius comes into play, drawing the scapula and clavicle up towards the occiput.) The front part of the deltoid muscle, used alone, helps to flex the shoulder, moving the humerus forwards, and the back portion, used alone, helps to extend the shoulder, moving the humerus backwards and towards the mid-line.

The *brachialis* lies lower in the front of the arm than the biceps, arising from the humerus and being inserted into the coronoid process of the ulna. It assists the biceps in the powerful action of flexion of the elbow (Table 14.4).

The Muscles of the Forearm

The forearm contains numerous small, less powerful muscles for movement of the wrist and digits (Fig. 14.9). In the front lie the flexors of the wrist, the common flexors of the fingers, the long flexor of the thumb and the pronator muscles of the wrist. The common flexors of the digits divide into four tendons, which run across the palm of the hand and up to the terminal phalanx of each digit, into which they are inserted. At the back lie the extensors of the wrist, the common extensors of the fingers, the extensor of the thumb and of the first finger, and the supinators of the wrist. The tendons of the muscles which cross the wrist are bound down by the *flexor retinaculum* just above the wrist joint. In the same way the

Table 14.4 Muscles of the Arm

Name	Position	Origin	Insertion	Action
Biceps (two heads)	The front of the arm	Coracoid process and above the glenoid cavity of the scapula	Radial tuberosity	Flexes elbow and shoulder and supinates the hand
Triceps (three heads)	The back of the arm	One head from the axillary border of the scapula; two from the shaft of the humerus	Olecranon of ulna	Extends elbow and shoulder
Deltoid	Over the shoulder	Acromion and spine of scapula; the clavicle	Deltoid tuberosity of humerus	Abducts the shoulder to a right angle
Brachialis	Crosses the front of the elbow	Humerus	Anterior surface of coronoid process of ulna	Flexes the elbow

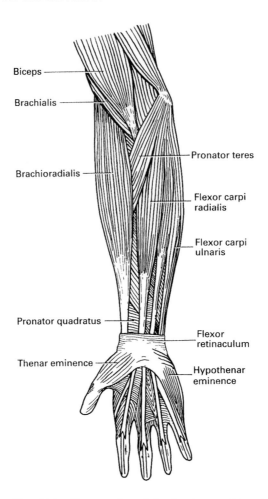

Biceps

Brachialis

Pronator teres

Brachioradialis

Flexor carpi radialis

Flexor carpi ulnaris

Pronator quadratus

Flexor retinaculum

Thenar eminence

Hypothenar eminence

Fig. 14.9 The muscles of the forearm and hand.

tendons are bound down to the fingers to keep them close to the bones.

The Muscles of the Hand

The hand contains very little muscle, as this would tend to make it clumsy and interfere with its usefulness in grasping and lifting. Many of the muscles which move it are therefore in the forearm (Fig. 14.9). The hand only contains the short flexor of the thumb, and the adductor and abductor muscles for the digits. These latter are termed the interosseous muscles, and are only well developed at the base of the thumb and to a lesser extent at the base of the little finger. Here they form the *thenar* and *hypothenar eminences,* and give power to the grip, where adduction of the thumb is particularly important.

The muscles of the lower limb are much larger and more powerful than those of the upper, as the limb carries the whole weight of the body. They may be divided into the muscles of the thigh, the leg and the foot.

The Muscles of the Thigh

The muscles of the thigh are particularly strong and include those listed in Box 4.6 and shown in Figs 14.10 and 14.11.

The muscles of the thigh

- The quadriceps femoris
- The hamstrings
- The sartorius
- The adductors of the hip

Box 14.6

The *quadriceps* is so called because it has four heads; rather it is four muscles with a common insertion into the patella, and through the patellar ligament it joins the tibia and is the extensor of the knee joint, used in standing and in the powerful action of kicking. It is made up of the *rectus* or straight muscle, and the three *vastus* muscles, the lateral, intermedial and medial. Of these the lateral is the longest muscle and lies on the outside of the thigh. It is occasionally used for the giving of intramuscular injections, as it is well away from the blood vessels, nerves and lymphatics of the limb here.

The *hamstrings* are the flexor muscles of the knee and are so called because of the strong tendons or 'strings' that they form on either side of the popliteal space, at the back of the knee joint. These can be seen and felt readily when the knee is bent. The hamstring muscles are (Fig. 14.12):

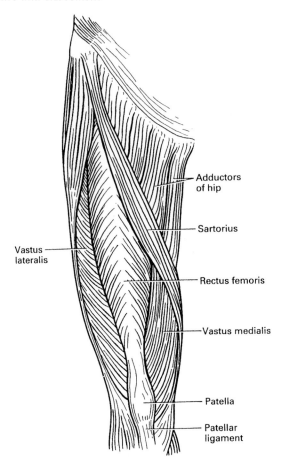

Fig. 14.10   The muscles of the front of the thigh.

1.   The *biceps femoris*, so called because it arises by two heads, one from the ischial tuberosity, the other from the back of the femur. It is inserted into the fibula. It lies on the outside of the back of the thigh.

2.   The *semitendinosus* is so called because of the length of the tendon by which it is inserted into the tibia. It arises from the ischial tuberosity, with the biceps muscle, and lies in the middle of the back of the thigh.

3.   The *semimembranosus* is so called because of the membranous tendon by which it arises from the ischial tuberosity. It is inserted into the tibia and lies on the inner side of the back of the thigh.

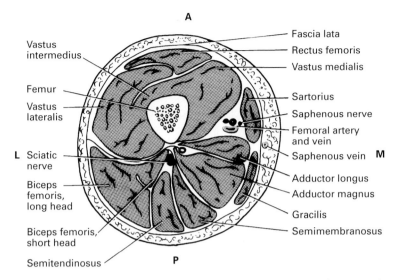

Fig. 14.11 Section through the thigh, showing the relation of the muscles to the bone and other structures. L, Lateral; M, medial; A, anterior; P, posterior.

The biceps thus forms the 'hamstring' tendon on the outer side of the back of the knee and the semitendinosus and the semimembranosus form the 'hamstring' tendon on the inner side of the back of the knee. The hamstring muscles are again a very powerful group of muscles serving to bend the knee in walking, jumping and climbing, and also, when the tibia is fixed, as in standing, pulling on the ischial tuberosity and helping the gluteal muscles to extend the hip joint.

The *sartorius* or tailor's muscle, runs from the superior anterior iliac spine across the front of the thigh to the inner side of the knee, which it crosses, and is inserted into the tibia. It assists in the joint movements which occur when sitting tailorwise, flexing hip and knee and rotating the femur.

The *adductor muscles* form the flesh on the inside of the thigh and are muscles by which the hip is adducted, being assisted by the smaller, more superficial muscle, the *gracilis*. These muscles are particularly well developed in persons who ride horses, since the rider holds on by the knees, keeping the hips adducted. The adductor muscles run from the pubis and ischium and are inserted into the linea aspera and medial condyle of the femur. The adductor magnus forms the greater part of this muscle and a canal runs through it which allows the main artery of the thigh to run from the

Gluteal muscles

Biceps femoris ⎫
Semitendinosus ⎬ Hamstrings
Semimembranosus ⎭

Adductors of the hip

Gastrocnemius

Soleus

Tendo calcaneous

Fig. 14.12　The muscles of the back and lower limbs.

inner side of the thigh to the back of the knee, where it is well protected (Table 14.5).

The Muscles of the Leg

In the leg there are a few large muscles controlling the ankle and many smaller ones moving the foot. The chief muscles are listed in Box 14.7 and shown in Figs. 14.13 and 14.14.

The *gastrocnemius* and *soleus* together form the flesh of the calf, the gastrocnemius lying at the back and the soleus lying in front of it. The *gastrocnemius* arises by two heads from the femur, forming the borders of the popliteal space below, as the hamstrings do above the knee. The *soleus* arises from the tibia, not crossing the

Table 14.5 Muscles of the Thigh

Name	Position	Origin	Insertion	Action
Quadriceps femoris	Front of thigh	Ilium and femur	Patella, through which it is joined to the tibia by the patellar ligament	Extension of knee and flexion of hip
Hamstrings	Back of thigh	Ischial tuberosity and femur	Tibia and fibula by tendons at either side of the popliteal space	Flexion of the knee and extension of the hip
Sartorius	Crosses front of thigh	Anterior, superior iliac spine	Tibia at inner side below the knee	Assists in flexion of hip and knee and abduction and external rotation of hip
Adductors of the hip	Inner side of thigh	Pubis and ischium	Linea aspera and medial condyle of femur	Adduct the hip

---

The muscles of the leg

- The gastrocnemius ⎫
- The soleus ⎬ the calf muscles
- The tibialis anterior
- The flexors and extensors of the digits

Box 14.7

---

knee joint and thus not affecting its movements. Both muscles unite below to form a strong common tendon, the tendo calcaneus, by which they are inserted into the calcaneum. The calf muscles raise the heel, causing plantar flexion, or, as it is sometimes called today, extension, of the ankle joint, as in walking and running.

The *tibialis* lies in front of the leg just outside the crest of the tibia, where it can be seen and felt when the toes and ball of the foot are raised from the ground. It is this muscle which gets stiff from abnormal exercise, when persons unused to walking up steep slopes begin to climb mountains. It arises from the tibia and fibula below the knee joint, and is inserted into the tarsal and metatarsal bones on the inside of the instep, where its tendon can be seen and felt readily when the ball of the foot is raised from the ground and the sole turned to face in towards the midline. It produces dorsi-

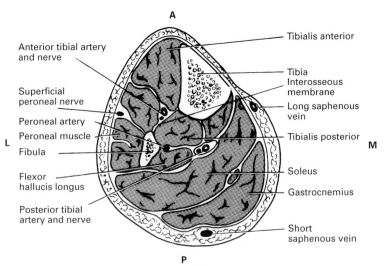

Fig. 14.13 Section through the leg, showing the position of the muscles in relation to the bones. A, Axis of the centre of the limb; M, median; L, lateral.

Tibialis anterior

Gastrocnemius

Soleus

Extensor retinacula

Medial malleolus

Lateral malleolus

Extensor tendons

Fig. 14.14 The muscles of the front of the leg and foot.

flexion of the ankle, i.e. bending towards the dorsal surface, or, as it is sometimes called today, flexion, bringing the terminology into line with that used for all other joints. It also inverts the foot (Table 14.6).

The Muscles of the Foot

The foot, like the hand, contains little muscle, the chief muscles moving the foot lying largely in the leg. The tendons of the

Table 14.6 Muscles of the Leg

Name	Position	Origin	Insertion	Action
Gastrocnemius	The calf	Condyles of the femur	Calcaneum	Plantar flexion of the ankle; raising the heel
Soleus	The calf	Tibia and fibula	Calcaneum	
Tibialis anterior	Side of the leg	Tibia	Tarsals and metatarsals inside the instep	Dorsiflexion of the ankle, raising the toes; inversion of the foot

extensors of the digits cross the dorsal surface of the foot, the big toe having an individual muscle and tendon. The tendons of the flexors of the digits cross the sole and are strong and very important in helping to support the arch of the foot. There is a common flexor for the small toes and a flexor for the big toe. In addition, a short flexor of the toes crosses the sole from the calcaneum to the phalanages, and also gives support to the arch. Small interosseous muscles between the metatarsal bones abduct and adduct the digits, but are little used, and therefore little developed.

SECTION IV: INTERNAL TRANSPORT

Section Introduction

Two closely related systems, in both the anatomical and physiological sense, are covered in this section. These systems are the cardiovascular system and the respiratory system and these, respectively, are responsible for circulating blood around the body and delivering oxygen to the blood and removing carbon dioxide. The lymphatic system, which also plays a key role in the defence of the body, is included here because it is responsible for removing excess fluid from the interstices of the peripheral tissues and returning it to the circulation.

The cardiovascular system is composed of the heart and the blood vessels and it is their collective action, with the heart working as a circulatory pump and the blood vessels working as a closed transporting system, which delivers blood to all the parts of the body. The blood is a tissue which has several functions, including transport of gases, defence against infection, clotting, distribution of heat and transport of nutrients. The gas transport and clotting functions are considered in this section.

Primarily, the respiratory system exchanges oxygen and carbon dioxide between atmospheric air and the blood. This is achieved mechanically in respiratory movements which draw in atmospheric air and expel air rich in carbon dioxide. Air is mixed in the alveoli to produce alveolar air, and gas exchange between the blood and alveolar air takes place here by diffusion across the thin walls of the pulmonary capillaries and the alveolar membranes. These mechanisms are described.

Chapter 15
The Blood

The circulatory system is the transport system of the body by which food, oxygen, water and all other essentials are carried to the tissue cells and their waste products are carried away. It consists of three parts (Box 15.1).

The circulatory system

- The **blood**—the fluid in which materials are carried to and from the tissues
- The **heart**—the driving force which propels the blood
- The **blood vessels**—the routes by which the blood travels to and through the tissues, and back to the heart

Box 15.1

The blood is a thick red fluid; it is bright red in the arteries, where it is oxygenated, and a dark purplish-red in the veins, where it is deoxygenated, having given up some of its oxygen to the tissues—the cause of the colour change—and received waste products taken in from them. It is slightly alkaline, and the pH varies very little during life, as the cells of the body can live only if the pH is normal. It forms about five per cent of the body weight, so that the average volume is three to four litres.

COMPOSITION OF THE BLOOD

Although apparently fluid, blood actually consists of a fluid and a solid part. When examined under the microscope, large numbers of small round bodies can be seen in it which are known as the *blood corpuscles* or cells. These form the solid part, while the liquid in which they float is the fluid part and is called plasma. The cells form 45 per cent and the plasma 55 per cent of the total volume.

Plasma

The plasma or fluid part of the blood is a clear, straw-coloured, watery fluid, similar to the fluid found in an ordinary blister (Box 15.2).

The components of plasma

- **Water:** which forms over 90 per cent of the whole
- **Mineral salts:** these include chlorides, phosphates and carbonates of sodium, potassium and calcium. The chief salt present is sodium chloride. The correct balance of the various salts is necessary for the normal functioning of the body tissues, and there is a total of 0·9 per cent inorganic substances
- **Plasma proteins:** albumin, globulin, fibrinogen, prothrombin and heparin
- **Foodstuffs in their simplest forms:** glucose, amino acids, fatty acids and glycerol, and vitamins
- **Gases in solution:** oxygen, carbon dioxide and nitrogen
- **Waste products from the tissues:** urea, uric acid and creatinine
- **Antibodies and antitoxins:** these protect the body against bacterial infection
- **Hormones:** from the ductless glands
- **Enzymes**

Box 15.2

The *water* in plasma provides fresh water to supply the fluid which bathes all the body cells and renews the water within the cells. Sixty per cent of the body weight is water and in an average man weighing 70 kilograms (kg) this would be approximately 46 litres. Of the 46 litres approximately 29 litres are within the cells (intracellular fluid) and 17 litres are outside the cells (extracellular fluid). The extracellular fluid is divided between the blood vessels (3 litres) and the fluid bathing the cells, called the interstitial fluid (14 litres).

The *salts* in the plasma are necessary for the building of protoplasm and they act as buffer substances neutralizing acids or alkalies in the body and maintaining the correct pH of the blood. Blood is always slightly alkaline in health and has a pH of 7·4 (see Chapter 1). In plasma there are approximately 155 mmol/litre of positively charged ions, chiefly sodium, balanced by 155 mmol/ litre of negatively charged ions, mainly chloride and bicarbonate. This is referred to as the *electrolyte balance*, and is similar in the interstitial fluid. In the intracellular fluid potassium replaces sodium as the positively charged ions, and phosphate ions and proteins replace chloride as the negatively charged ions.

The *proteins* which plasma contains give the blood the sticky consistency, called viscosity, which is necessary to prevent too much fluid passing through the capillary walls into the tissues. If there is a deficiency of protein as in kidney disease when protein is constantly lost as albumin in the urine, the osmotic pressure of the plasma is lowered and excess fluid escapes into the tissues. This excess fluid in the tissues is called oedema. The viscosity of the blood also assists in the maintenance of the blood pressure. Albumin is thought to be formed in the liver and globulin is derived from the group of white blood cells called lymphocytes. Fibrinogen and prothrombin are produced in the liver and are both necessary for the mechanism of the clotting of blood. Plasma without fibrinogen is called serum; this can be seen as the yellow fluid which oozes from a cut after a clot has formed. Heparin is also formed in the liver and prevents blood clotting in the vessels.

Foodstuffs, in the form of glucose, amino acids, fatty acids and glycerol, are absorbed from the alimentary tract into the blood. They are the end-products of carbohydrate, protein and fat metabolism.

Urea, uric acid and creatinine are the *waste products* from protein metabolism. They are made in the liver and are carried by the blood for excretion by the kidneys.

Antibodies and antitoxins are complex protein substances which provide protection against infection and neutralize the poisonous bacterial toxins.

Enzymes are chemical substances produced by the body, which produce chemical changes in other substances without themselves entering into the reaction.

The Blood Cells

The cells are of three types: red blood cells (erythrocytes), white blood cells (leucocytes) and platelets (thrombocytes).

The Formation of Blood Cells
The formation of blood cells takes place in the bone marrow and the mature products are released into the blood stream. Eight

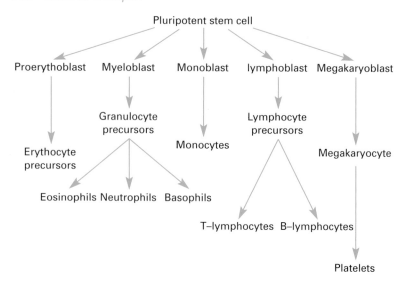

Fig. 15.1   Summary of the formation of blood cells.

different cells are formed and all of these are formed from one type of *pluripotent stem cell* which gives rise to five different lines of cells (Fig. 15.1). The *myeloblast* line gives rise to three types of *granulocyte cells* and the *monoblast* and *lymphoblast* lines give rise to the *agranulocyte cells*. The *erythrocytes* (red cells) and the *platelets* are formed from their own specific cell lines.

Red Cells

The red cells are minute disc-shaped bodies, concave on either side (Fig. 15.2). They are very numerous, numbering about 5 000 000 per cubic millimetre of blood. They are very minute, having a diameter of 7·2 micrometres only (1 micrometre = 1/1000 millimetre; this can be abbreviated to µm). They have no nucleus, but contain a special protein known as *haemoglobin*. This is a pigment and is yellow in colour, though the massed effect of these numerous yellow bodies is to make the blood red. Haemoglobin contains a little iron, and this iron is essential to normal health, though the total amount in the whole body is said to be only sufficient to make a two-inch nail. Haemoglobin has a great attraction for oxygen. As the red cells pass through the lungs the haemoglobin combines with oxygen from the air (forming oxyhaemoglobin) and becomes bright in colour. This makes the oxygenated blood bright red. As the red cells pass through the tissues oxygen is given off from the blood and the haemoglobin becomes a

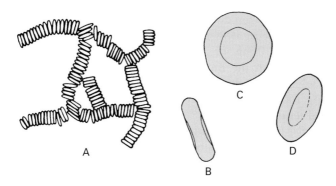

Fig. 15.2   Red cells. (A) In rouleaux, as seen in shed blood under the microscope. (B–D) Three views of a single cell.

dull colour (reduced haemoglobin) making the blood a dark purplish-red. The haemoglobin is measured in grams per 100 ml; the normal figure is 14 to 16 g per 100 ml.

The *function* of the red cells is therefore to carry oxygen to the tissues from the lungs and to carry away some carbon dioxide. This is their sole function, and is dependent on the amount of haemoglobin they contain. If there is a lack of haemoglobin, either because red cells are reduced in number or because each one does not contain the normal quantity of haemoglobin, the individual suffers from anaemia.

The red cells are *produced* in the red bone marrow of spongy bone, which is found in the extremities of long bones and in flat and irregular bones. In childhood the red bone marrow also extends throughout the shaft of the long bones, as children have greater need for the production of red cells.

Red cells pass through various stages of development in the bone marrow. *Erythroblasts* are large cells containing nuclei and a small quantity of haemoglobin. These develop into *normoblasts* which are smaller cells with more haemoglobin and smaller nuclei. The nucleus then disintegrates and disappears and the cytoplasm contains fine threads, at which stage the cells are called *reticulocytes*. Finally the threads disappear and the fully mature *erythrocyte* passes into the blood stream. In health almost all red cells in the blood should be erythrocytes, with only an occasional reticulocyte. Many factors are necessary for the normal formation of red blood cells.

Protein is required for the manufacture of protoplasm.

Iron is needed for the haemoglobin. Very little iron is excreted. As the red cells are broken down the iron is stored and used again, but a certain amount of iron must be taken in in the diet. A man requires about 10 mg of iron per day, women require about 15 mg to replenish the menstrual loss and the depletion of iron reserves which occur during pregnancy, labour and lactation. Iron-containing foods are red meat, egg-yolk, green vegetables, peas, beans and lentils.

Vitamin B₁₂ (cyanocobalamin) is necessary for the maturation of red blood cells and is usually found in adequate quantities in the diet in temperate climates. It can be absorbed from the small intestine only when it has been combined with the intrinsic factor which is secreted by the stomach. Together these two substances are known as the anti-anaemic factor (or haemopoietic factor), which is stored in the liver and passed to the bone marrow as necessary. Vitamin B_{12} is also known as the extrinsic factor.

Other factors which are necessary even though in small quantities are vitamin C, folic acid (one of the vitamin B complex), the hormone thyroxine and traces of copper and manganese.

Red blood cells live in the circulation for about 120 days after which they are ingested by the cells of the monocyte/macrophage system in the spleen and lymph nodes. Here the haemoglobin is broken down into its component parts, which are carried to the liver. The globin is returned to the protein stores or is excreted in the urine after further breaking down. The haem is further split into iron, which is stored and used again, and pigment, which is converted by the liver into bile pigments and is excreted in the faeces. Red cell production and breakdown usually proceed at the same rate so the number of cells remains constant.

White Cells
The white cells, or *leucocytes*, are larger than the red cells, measuring about 10 μm in diameter, and they are less numerous. There are $7–10 \times 10^9$ per litre of blood, though this number increases considerably to 30×10^9 per litre when infection is present in the body. This increase is known as *leucocytosis*. The leucocytes are of three different types (Fig. 15.3):

1. *Polymorphonuclear leucocytes* are also known as *granulocytes* because of the granular appearance of the cytoplasm. The nucleus gradually develops several lobes, hence the name (poly = many, morph = form). These cells make up about 75 per cent of the total white cells. They are made in the red bone marrow and

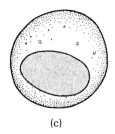

(a)

(b)

(c)

Fig. 15.3 White cells. (A) A polymorphonuclear leucocyte. (B) A lymphocyte. (C) A monocyte.

survive for about 21 days. Granulocytes can be further classified according to their staining properties. Neutrophils (70 per cent) absorb both acid and alkaline dyes. They are able to ingest small particles, e.g. bacteria and cell debris. This ability is called phagocytosis and they are sometimes known as phagocytes. They have amoeboid movement and can pass out of the blood stream through the capillary walls to accumulate where there is infection.

Eosinophils (four per cent) absorb acid dyes and stain red. An increase in their number occurs during allergic states such as asthma, and during infestation with worms.

Basophils (one per cent or less) absorb alkaline or basic dyes and stain blue. They contain heparin and histamine.

2. *Lymphocytes* make up about 20 per cent of the total white cell count. They are made in the lymph nodes and in the lymphatic tissue which is present in the spleen, the liver and other organs. They show some amoeboid movement but are not actively phagocytic. They are concerned in the production of antibodies.

3. *Monocytes* make up about five per cent of the total white cell count. They are the largest of the white blood cells and have a horse-shoe-shaped nucleus. They show both amoeboid movement and are phagocytic and are part of the reticulo-endothelial system (see p.348).

Platelets

Platelets, or *thrombocytes*, are even smaller than red blood cells and are made in the bone marrow. There are about 250×10^9/litre of blood. They are necessary for the clotting of blood.

THE CLOTTING OF BLOOD

The process whereby blood loss from the body is prevented following a cut is called *haemostasis* and it involves three stages which work together (Box 15.3).

The stages of haemostasis

- **Vascular spasm**—narrowing of the lumen of the cut blood vessels to slow down the loss of blood
- **Formation of a platelet plug**—to stop the loss flow of blood from the cut
- **Clotting of fibrin around the plug and retraction of fibrin**—to seal the cut and pull the edges of the cut together

Box 15.3

The clotting process (Fig. 15.4) is very complex and involves many factors. The end point in the process is the formation of an insoluble *fibrin clot* from soluble *fibrinogen* and this process is stimulated by the formation of *thrombin*. The formation of thrombin is stimulated by the formation of *prothrombin activator* and there are two systems whereby this is achieved—the *extrinsic* and the *intrinsic systems*. The extrinsic system is stimulated by damaged tissue and quickly forms a very small amount of fibrin to form a clot. The intrinsic system takes a few minutes to work but leads to the formation of a relatively large amount of fibrin to complete the formation of the clot. As soon as the clot is formed it is broken down by the action of an enzyme called *plasmin*. This allows the clot to be removed in order to begin the process of wound healing (Chapter 25).

Factors Affecting Clotting

Prothrombin is made in the liver and vitamin K is necessary for its manufacture. Vitamin K is present in green vegetables and is also manufactured in the intestines by bacterial action. It can be absorbed from the intestines into the blood only in the presence of bile. If bile is not present, as in some forms of jaundice, prothrombin may be lacking and the tendency to bleed is increased.

Heparin is a protein normally present in blood, which is formed in the liver and which prevents blood clotting in the vessels. It is called an anticoagulant.

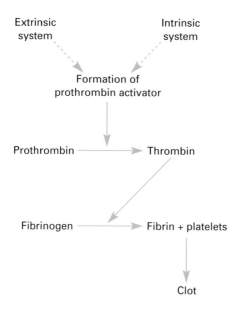

Fig. 15.4 The clotting process.

THE FUNCTIONS OF BLOOD

The uses of blood are listed in Box 15.4.

Functions of blood

- To carry nutrients to the tissues
- To carry oxygen to the tissues in oxyhaemoglobin
- To carry water to the tissues
- To carry away waste products to the organs which excrete them
- To fight bacterial infection through the white cells and antibodies
- To provide the materials from which glands make their secretions
- To distribute the secretions of ductless glands, and enzymes
- To distribute heat evenly throughout the body, and so regulate the body temperature
- To arrest haemorrhage through clotting

Box 15.4

BLOOD GROUPS

Blood from one individual cannot always be safely mixed with that of another. This fact became evident with the introduction of blood transfusion, which at first sometimes cured but sometimes killed the patient. This was due to the fact that blood is of four basic groups. If blood of differing groups is mixed, the red corpuscles may become sticky and form clumps. This is termed *agglutination*, and is fatal as the clumps of red cells block the blood vessels and obstruct the circulation, and the kidneys are severely damaged by excretion of excessive quantities of pigment from destroyed red cells.

When a blood transfusion is required it is necessary first to find which group the person belongs to and then to find a donor belonging to the same group. Blood groups are named according to the presence or absence of substances called *agglutinogens*, which are present in the red blood cells. There are two agglutinogens, called A and B. If A is present the blood group is called A, if B is present the blood group is called Group B, if both A and B are present the blood group is called Group AB, and if neither agglutinogen is present the group is called O. Having found to which group the patient belongs (Table 15.1) and found blood donated by a person belonging to the same basic group, a sample of red blood cells from the donor's blood is mixed with some plasma from the patient who is to receive the transfusion (now called the recipient). This is because plasma contains substances called *agglutinins* which cause agglutination of red cells if incompatible blood groups are mixed. Agglutinins are called anti-A and anti-B, and plasma contains all agglutinins which will not affect its own red cells. Therefore the plasma of Group A contains anti-B agglutinin, the plasma of Group B contains anti-A agglutinin, the plasma of group AB contains no agglutinins, and the plasma of Group O contains both anti-A and anti-B agglutinins. When the donor's red cells are mixed

Table 15.1 Blood Groups

Blood group	Agglutinogen in red cells	Agglutinin in plasma	Transfusion possible
A	A	Anti-B	Groups A and O
B	B	Anti-A	Groups B and O
AB	A and B	Neither	Any group
O	None	Anti-A and anti-B	Group O only

with the recipient's plasma in the laboratory, it can be seen with the help of a microscope whether agglutination occurs. It will be noticed that Group AB has no agglutinins in the plasma and therefore cannot cause any red cells to agglutinate; this means that a patient with blood belonging to this group will probably be able to receive blood from any other group and the group is therefore known as the universal recipient. Group O contains no agglutinogens in the red cells and therefore they cannot be made to agglutinate by the agglutinins in any plasma. Blood belonging to this group can therefore be given to a patient belonging to any group and it is known as the universal donor. In practice the compatibility of blood is always checked very carefully before it is given to a patient.

Rhesus Factor

In addition to ABO grouping there is an additional factor present in the blood of about 85 per cent of the population. It is an agglutinogen called the Rhesus factor. Those who possess the factor are called Rhesus positive (Rh+), the remaining 15 per cent are called Rhesus negative (Rh−). If an Rh− person receives the blood of an Rh+ donor the agglutinogen stimulates the production of anti-Rh agglutinins called anti-D. If a second Rh+ transfusion were given later the transfused cells would be agglutinated and destroyed (haemolysed) with serious or fatal results to the recipient. This factor can also cause difficulty during pregnancy. If an Rh− mother is carrying an Rh+ fetus the mother may begin to produce anti-Rh agglutinins which may then destroy the baby's red cells. The baby may overcome this spontaneously, or may require exchange transfusion.

Chapter 16
The Heart and Blood Vessels

The purpose of this chapter is to consider the anatomical location and structure of the heart and the major blood vessels. Some aspects of the function of the heart are included such as the cardiac cycle, the pathway of blood around the heart and the generation of electrical activity at the sino-atrial node.

THE HEART

The heart is a hollow, muscular, cone-shaped organ (Fig. 16.1). It lies between the lungs in an area called the *mediastinum*, behind the

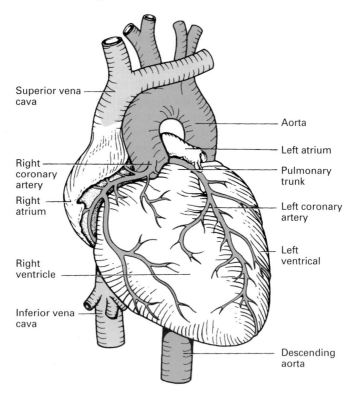

Fig. 16.1 Front view of the heart.

body of the sternum with two-thirds of its bulk on the left side. The circular *base* of the cone is directed upwards and to the right and the *apex* points downwards, forwards and to the left. It is usually on the level of the fifth intercostal space about 9 cm from the midline. The heart measures about 12 cm from base to apex, about 9 cm in width and is about 6 cm thick.

General Structure of the Heart

The heart is divided from base to apex by a muscular partition called the *septum*. The two sides of the heart have no communication with each other in health. Additionally each side is subdivided into an upper and lower chamber. The upper chamber on each side, the *atrium*, is smaller and is a receiving chamber into which the blood flows through veins. The lower chamber, the *ventricle*, is the discharging chamber from which the blood is driven into the arteries. Each atrium communicates with the ventricle below it on the same side of the heart through an opening, guarded by a valve called the *atrio-ventricular valve* (Fig. 16.2).

The heart is composed of cardiac muscle, the *myocardium*, on the action of which the circulation of the blood depends. The myocar-

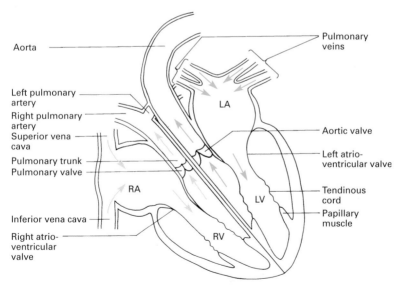

Fig. 16.2 Diagram showing the flow of blood through the heart. L.A., Left atrium; L.V., left ventricle, R.A., right atrium; R.V., right ventricle.

dium varies in thickness, being thickest in the left ventricle, thinner in the right ventricle and thinnest in the atria.

The atria and ventricles are lined with a thin, smooth, glistening membrane called the *endocardium* which consists of a single layer of endothelial cells and which is continuous with the valves and with the lining of the blood vessels.

The *pericardium* covers the heart and the roots of the great vessels and has two layers. The outer layer or *fibrous pericardium* is securely anchored to the diaphragm, the outer coats of the great vessels and the posterior surface of the sternum and therefore maintains the heart in its position in the chest. Because of its fibrous nature it also prevents overdistension of the heart. The inner layer, the *serous pericardium*, lines the fibrous pericardium and is invaginated by the heart; it therefore also has two layers. The inner layer is known as the visceral portion, or *epicardium*, and it is reflected back to form the outer or parietal layer. The layers are normally all in close contact and are moistened by fluid which exudes from the serous membrane; this prevents any friction as the heart continually contracts and relaxes. In inflammatory conditions, such as pericarditis, the amount of fluid within the pericardium may embarrass the action of the heart and may be aspirated.

Valves of the Heart
The heart is provided with *valves* to prevent the blood from flowing in the wrong direction. There are four main valves.

The *right atrio-ventricular* (or *tricuspid*) *valve* lies between the right

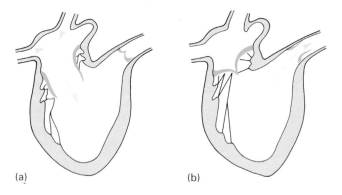

(a)   (b)

Fig. 16.3   The action of the valves of the heart. (a) The position of the valves during contraction of the atrium and relaxation of the ventricle. (b) The position of the valves during contraction of the ventricle and relaxation of the atrium.

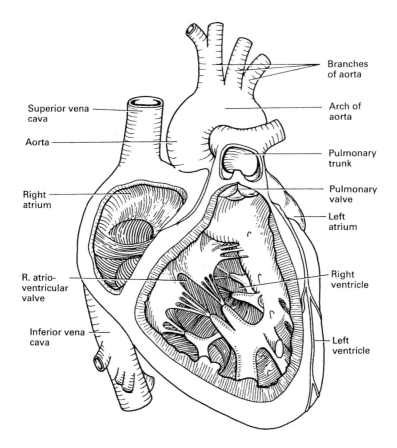

Fig. 16.4 Inside the right side of the heart.

atrium and the right ventricle. It consists of three triangular flaps or cusps, each consisting of a double layer of endocardium strengthened with fibrous tissue. The under surface of the cusps gives attachment to a number of fine, tendinous cords, called *chordae tendinae*, which originate in the *papillary muscles* in the wall of the ventricle. When the ventricle contracts the blood is pushed back towards the atrio-ventricular opening but is prevented from entering the atrium by the cusps of the valve, which close because of the increased pressure in the ventricle. The contracton of the papillary muscles exerts tension on the chordae tendinae which prevent the valve cusps from being carried into the right atrium (Figs 16.3 and 16.4).

The *left atrio-ventricular valve* is also called the *mitral valve* because it has only two cusps (Fig. 16.5). The structure is similar

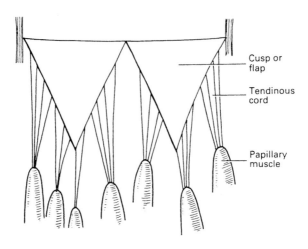

Fig. 16.5 Diagrammatic representation of the mitral valve laid open.

to that of the right atrio-ventricular valve. It prevents the backflow of blood into the left atrium during contraction of the left ventricle.

The *aortic valve* consists of three cusps which surround the entrance into the aorta from the left ventricle (Figs 16.6 and 16.7). The cusps are half-moon shaped and are fixed by their curved edges to the wall of the aorta, the straight edge being free so that pockets are formed facing into the aorta. As the blood flows from the left ventricle into the aorta the cusps lie flat against the vessel wall; as the ventricle relaxes the pockets fill with blood and bulge out, meeting in the centre and blocking the opening completely thus preventing blood flowing back into the ventricle. The *coronary arteries*, which supply the heart muscle with oxygenated blood, arise from the aorta just above the attached edges of the cusps of the aortic valve.

The *pulmonary valve* guards the opening from the right ventricle into the pulmonary trunk. It is similar in structure and action to the aortic valve.

The blood which returns to the heart from the myocardium passes through the *coronary sinus* and pours directly into the right atrium. The opening of the coronary sinus is protected by a thin, semi-circular valve called the *valve of the coronary sinus* which prevents backflow of blood into the sinus during contraction of the right atrium. There is also an imperfect valve guarding the opening from the inferior vena cava into the right atrium; it is called the *valve of the inferior vena cava.*

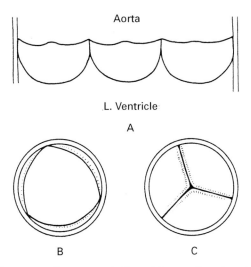

Fig. 16.6 Diagrammatic representation of the aortic valve. (A) Laid open. (B) Seen from above. (C) Closed.

The blood supply to the heart muscle is by the *right and left coronary arteries* which are the first branches from the aorta. There are a number of places within the heart muscle where these two arteries join, or anastomose, but most of the blood returns from the myocardium in veins which empty into the *coronary sinus*.

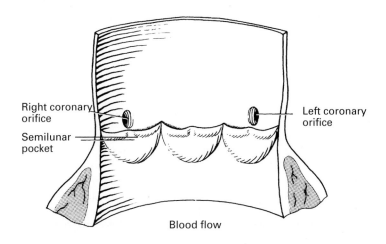

Fig. 16.7 The aortic valve.

The Function of the Heart

The heart is a pump whose purpose is to drive the blood into and through the arteries but the right and left sides of the heart function quite separately from one another.

Blood from all parts of the body is returned to the *right atrium* through the two large veins, the *superior* and *inferior venae cavae*. When it is full the right atrium contracts and drives the blood through the right atrio-ventricular valve into the *right ventricle* which in turn contracts sending the blood through the *pulmonary valve* and into the *pulmonary trunk*. The pulmonary trunk divides into *right and left pulmonary arteries* which carry the blood to the lungs where the gaseous exchange occurs. The blood is finally collected up into four *pulmonary veins* which return the blood to the *left atrium*. When it is full the left atrium contracts, simulta-neously with the right atrium, and the blood is driven through the left atrio-ventricular valve into the *left ventricle*. This chamber contracts, simultaneously with the right ventricle, and sends the blood into the *aorta*, the main artery of the body.

The heart contracts about seventy to eighty times each minute throughout life, though the rate varies with age, emotion, exercise

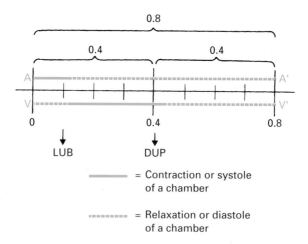

Fig. 16.8 Graph to show the sequence of events in the cardiac cycle. A–A′, the contraction and relaxation of the atrium; V–V′, the contraction and relaxation of the ventricle. The cycle occupies 0·8 second, of which 0·4 second is occupied by the contraction of both atria and ventricles, and the remaining 0·4 second by relaxation of all chambers of the heart. (Note: 0·8 second is the time of one beat if the pulse rate is 75 per minute.)

and other influences. Each beat is a cycle of events which lasts about 0·8 second (Fig. 16.8).

Blood pours into the atria from the great veins until both are full and they then contract simultaneously, emptying their contents into the ventricles. Atrial contraction lasts about 0·1 second. Rising pressure in the ventricle as blood enters forces the atrio-ventricular valves to close and causes the first heart sound, which can be heard through a stethoscope placed over the apex of the heart and has been likened to the word 'lub'. Ventricular contraction follows, lasting 0·3 second, and forcing open the pulmonary and aortic valves. Blood is forced into the aorta and pulmonary trunk and as the ventricles relax the pressure in the great vessels forces the aortic and pulmonary valves to close and causes the second heart sound, which can be heard best over the second right rib and has been likened to the word 'dub' because it is a sharper sound. During ventricular contraction the atria are relaxed. Following ventricular contraction the whole heart is relaxed for approximately 0·4 second. The word for contraction is *systole* and that for relaxation is *diastole*.

The Conducting Mechanism of the Heart

Cardiac muscle, unlike skeletal muscle, has the property of being able to contract rhythmically independent of any nerve supply. The autonomic nervous system modifies the speed of the heart beat but the heart could be removed from an animal and continue to beat. The impulse to contract arises through spontaneous depolarization in an area of specialized tissue near the entry of the superior vena cava into the right atrium—called the *sinu-atrial node*—the pacemaker of the heart. The impulse then passes across both atria in more or less concentric rings, made possible by the fact that cardiac muscle fibres are branching, to the *atrio-ventricular node* situated in the septum near the junction of the atria and ventricles. After a slight pause the impulse spreads down the *atrio-ventricular bundle* which is in two strands, one in the right and one in the left ventricle. The strands break up into special fibres, called *Purkinje fibres*, from which branches pass beneath the endocardium to all parts of the ventricles.

The sinu-atrial node has the fastest rhythm, about 70 to 74 beats per minute and it imposes this rate on the other areas of conduction. The ventricles are able to contract independently of the atria and they will do this if the conducting mechanism is affected by disease, but they will contract much more slowly, about 40 beats per minute. This condition is known as *heart block* and may be serious as the tissues are unlikely to receive an adequate blood supply. In other cases some impulses pass down under the atrio-ventricular bundles but some do not so the ventricles contract once to every two or three atrial contractions; the condition is known as partial heart block.

The Electrocardiogram

If electrodes are placed on the body and attached to an instrument called an *electrocardiograph* it is possible to detect the electrical activity in the heart. The resulting pattern, which can be observed on an oscilloscope or traced onto paper is known as the *electrocardiogram*. This trace is helpful to physicians in allowing the detection of abnormalities, such as *myocardial infarction*, which alter the electrical activity in the heart. The trace is characteristic and has three phases (Fig. 16.9). The *P wave* represents the depolarization of the atria and the *QRS complex* which masks the *repolarization* (the return to the resting potential difference across the heart cell membranes) represents the depolarization of the ventricles. Finally, the *T wave* represents the repolarization of the ventricles.

Nervous Control of the Heart

The heart is innervated by the autonomic nervous system. The *vagus nerve* (tenth cranial nerve) slows the heart rate and causes decreased power of contraction by conveying impulses to the sinu-atrial node. The *sympathetic nerves* speed the heart rate and increase the force of contraction. (see also Chapter 5.) This dual innervation

P = Atrial depolarization
QRS = Ventricular depolarization
T = Ventricular repolarization
P – R interval = 0.2 seconds

Fig. 16.9   The electrocardiogram.

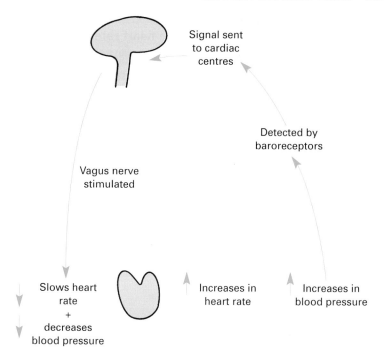

Fig. 16.10 Negative feedback loop controlling heart rate and blood pressure.

of the heart is coordinated by the cardiac centre in the medulla oblongata in the brain.

The heart rate is also controlled reflexly by two sets of receptors. *Pressure receptors* (or *baroreceptors*) are sensitive to changes in blood pressure. They are found in the carotid arteries and the arch of the aorta. If the blood pressure increases less sympathetic stimulation and more parasympathetic stimulation is observed and the heart rate slows, thus helping to lower the blood pressure. This is an example of a homeostatic mechanism which works by negative feedback (Fig. 16.10). *Chemoreceptors* are sensitive to the amount of oxygen and carbon dioxide present in the blood. They are found in the neck near the carotid arteries and near the aorta and are sensitive to lack of oxygen. Impulses are conveyed to the cardiac centre and the heart rate is accelerated to increase the blood supply, and therefore the oxygen supply, to the tissues.

Factors that influence heart rate

- Rest slows the heart rate and exercise quickens it
- Increasing age slows the heart rate; infants have a pulse rate of 120 to 140 at birth and it slows throughout life and into old age
- Women have a slightly faster pulse rate than men
- Emotion and excitement speed the pulse rate

Box 16.1

In health the heart rate varies considerably (Box 16.1). In disease, conditions such as fever, haemorrhage, shock and hyperthyroidism speed the heart rate, while increased pressure on the brain and heart block slow it. Drugs can also be used to speed or slow the pulse rate.

THE BLOOD VESSELS

Contraction of the ventricles sends the blood to all parts of the body through a complicated series of tubes called *arteries* which branch into tiny vessels called *arterioles*. These in turn are continuous with a network of microscopic vessels named *capillaries*. Blood is then collected into tiny vessels called *venules* which unite to form *veins*. These join one another forming large veins which finally return the blood to the heart.

Structure of the Blood Vessels

Arteries are thick-walled vessels and with one exception they carry oxygenated blood. The exception is the pulmonary trunk, which divides into two pulmonary arteries, and which carries deoxygenated blood from the right ventricle to the lungs. All arteries have three layers (Box 16.2 and Fig. 16.11).

All parts of the body must have a blood supply and the arteries are no exception. Very tiny blood vessels bring a blood supply to the arterial walls and equally small venules return the blood to the veins. Lymph vessels and nerve fibres are also present.

Arterioles have the same three structures as arteries, but the intima and media are thin and the adventitia is relatively thicker than the adventitia of an artery. There are also more muscle fibres and less elastic fibres.

The layers of arteries

- The outer coat, or *tunica adventitia*, which consists of collagenous and elastic fibres
- The middle coat, or *tunica media*, which consists mainly of smooth muscle with elastic fibres and some collagen fibres
- The lining, or *tunica intima*, which consists of a layer of endothelial cells and provides a smooth surface over which the blood can flow without clotting

Box 16.2

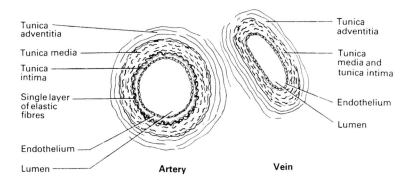

Tunica adventitia

Tunica media

Tunica intima

Single layer of elastic fibres

Endothelium

Lumen

Artery

Tunica adventitia

Tunica media and tunica intima

Endothelium

Lumen

Vein

Fig. 16.11   The structure of arteries (left) and veins (right).

Capillaries form a network between arterioles and venules (Fig. 16.12). They consist of a single layer of endothelial cells, similar to those which line all other blood vessels (Fig. 16.13).

Venules and *veins* both have three coats, as do arteries, but the

Arteriole

Venule

Fig. 16.12   Diagram of a capillary network.

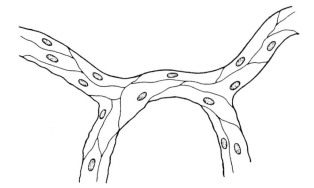

Fig. 16.13   Capillary network, showing the wall of simple endothelium.

middle coat of a vein is much thinner than that of an artery. Many veins are provided with *valves* to prevent blood flowing in the wrong direction. Each valve is formed of a double layer of endothelium strengthened by connective tissue and elastic fibres. The valve cusps are semi-lunar and are attached by their convex edges to the wall of the vein (Fig. 16.14). Veins, like arteries, have their own blood supply and nerve fibres though the latter are not so numerous.

THE MECHANISM OF CIRCULATION

The Pulse

Blood leaving the left ventricle of the heart is rich in oxygen and bright red in colour. It is pumped into the aorta by contraction of

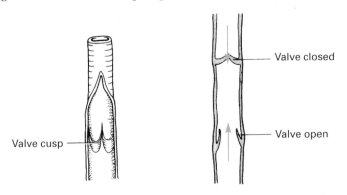

Fig. 16.14   Valves in the veins.

the left ventricle which initiates an area of increased pressure which travels along the arteries rather like a wave. When the blood is pumped out of the left ventricle the aorta is already full, so it must distend in order to accommodate the additional blood. As the left ventricle relaxes the aortic valve closes and the elastic aorta recoils to its original diameter. This recoil of the aorta is very important because it is the mechanism whereby the blood is continuously driven round the body even when the ventricle is relaxed. The distension and recoil of the aorta sets up a wave of distension and recoil called the *pulse* which travels along all the large arteries and which can be felt with the fingers wherever an artery can be compressed gently against a bone. Since the heart beat produces the pulsation the rate and character of the beat can be judged by taking the pulse.

Blood Pressure

The blood pressure is the force which the blood exerts on the walls of the blood vessels. It varies in the different blood vessels and also with the heart beat. The pressure is greatest in the large arteries leaving the heart, and gradually falls in the arterioles until, when it reaches the capillaries, it is so slight that the least pressure from without will obliterate these vessels and drive the blood out of them. This can be seen by pressing lightly on the nail or letting a piece of glass rest lightly on the skin. (It is for this reason that it is so important to change frequently the position of a patient confined to bed, as the tissue carrying the body's weight has little blood circulating through it.) In the veins the pressure is lower still until ultimately in the big veins approaching the heart there is suction, i.e. a negative pressure instead of a positive one, on account of the suction exerted by the heart as its chambers relax.

The pressure in the large arteries varies with the heart beat. It is highest when the ventricle contracts (this is called *systolic pressure*) and lowest when the ventricle relaxes (this is called *diastolic pressure*).

Measurement of Blood Pressure

The pressure of the blood is measured by the weight of the column of mercury which it will support. The height is calculated in millimetres. The normal arterial pressure is 110 to 120 millimetres systolic pressure and 65 to 75 millimetres diastolic pressure. The apparatus for measuring blood pressure is called a sphygmomanometer (Fig. 16.15). There are different types, but the most common consists of a hollow rubber cuff encased in material which is fixed

Fig. 16.15   (A) Diagnostic and (B) anaeroid sphygmomanometers.

evenly round the limb without constricting it in any way. To this is
attached a small hand pump, by which air can be pumped into the
cuff, expanding it and compressing the limb and the blood vessels
it contains. The air cushion is also linked to a manometer containing mercury, so that the pressure of the air within it can be read
against a graduated scale.

   To take the blood pressure the air cuff is fixed round the arm and
the pulse at the wrist is felt with one hand while the other is used to
inflate the cuff to a pressure above the point at which the radial
pulse can no longer be felt. A stethoscope is placed over the
brachial artery pulse in the cubital fossa and the pressure in the
cuff is slowly lowered by means of the release valve. As the
pressure falls no sounds are heard until the systolic blood pressure
is reached, at which point a tapping sound is heard in the stethoscope. At this point the level of mercury in the manometer must be
noted. As the pressure in the cuff is further reduced the sounds
become louder, until the diastolic blood pressure is reached, at
which point the sound is altered in character and becomes
muffled. A little further decrease in cuff pressure causes the sound
to disappear completely. The diastolic blood pressure is noted as
the point at which the character of the sound changes.

Arterial Pressure

The arterial pressure is maintained by five factors (Box 16.3).

Maintenance of arterial pressure

- The cardiac output
- The peripheral resistance
- The total blood volume
- The viscosity of the blood
- The elasticity of the arterial walls

Box 16.3

The *cardiac output* is the amount of blood pumped out per minute. When the left ventricle contracts, about 70 ml of blood is pushed into the aorta, which is already full, causing it to distend. This is the *stroke volume.*

The *peripheral resistance* is the resistance offered by the small blood vessels, particularly the arterioles, to the flow of blood. This resistance prevents blood flowing too quickly into the capillaries and in this way helps to maintain the blood pressure in the arteries. The lumen of the arterioles can be altered by the action of vaso-motor nerves and by adrenaline and noradrenaline from the adrenal glands. If the lumen is narrowed the resistance to blood flow is increased and arterial blood pressure is raised. If the lumen is enlarged more blood will pass through quickly to the capillaries and the blood pressure will be lowered.

The *blood volume* is the total amount of circulating blood. If this is reduced by the loss of whole blood, as in haemorrhage, or by loss of fluid from the circulation, as in shock, burns or dehydration, the blood pressure will be lowered.

The *viscosity of the blood* is its stickiness. Blood is two or three times more viscous than water. The viscosity depends partly on the plasma, particularly the plasma proteins, and partly on the number of red cells. The viscosity of the blood will be reduced if large quantities of intravenous normal saline are given, and lowered viscosity is associated with lowered blood pressure.

The *elasticity of the arterial walls* allows distension of the aorta when the ventricle contracts and elastic recoil as the ventricle relaxes. This recoil pushes the blood on and maintains the diastolic pressure of the blood. Distension and recoil occur throughout the arterial system. Elasticity may be reduced in atheroma which is a degenerative disease of the arteries. In this condition the blood pressure will be raised because of loss of elasticity.

High Blood Pressure
A high blood pressure is a systolic pressure of 150 to 180 mm and over. The diastolic pressure is also raised as a general rule, and a high diastolic pressure, e.g. 90 to 120 mm or more, is injurious, as it is a strain on the heart. A high blood pressure is dangerous in that the raised pressure may cause rupture of a blood vessel. Rupture is particularly likely to occur in the brain and is one of the causes of cerebrovascular accident or stroke. There is also a strain on the heart which may result in heart failure.

Low Blood Pressure
A low blood pressure is a systolic pressure of 100 mm or less. It occurs in cases of haemorrhage, shock and collapse, heart failure, and in disease of the suprarenal glands. The danger lies in an insufficient supply of blood to the vital centres of the brain. The treatment is therefore:

1.   To place the patient flat with the foot of the bed raised if necessary, to send blood by gravity to the vital centres of the brain.
2.   To give heart stimulants, e.g. adrenaline or noradrenaline, and circulatory stimulants such as adrenaline and noradrenaline to contract the blood vessels.
3.   To give fluids such as saline by intravenous, subcutaneous or rectal infusion, or blood or plasma transfusion, to increase the fluid in circulation.

As already mentioned, blood vessels have their own nervous supply. Arterioles are supplied with *vasomotor nerves* through which they can be dilated or contracted according to the needs of the tissues. The muscular walls of the arterioles are also affected by hormones from the adrenal glands, particularly adrenaline and noradrenaline, which cause contraction of the muscle fibres and consequent narrowing of the lumen of the arteriole. Arterioles do not pulsate but can dilate to carry more blood to an organ which is working and can contract so that they carry less blood when an organ is resting. In this way arterioles control the distribution of blood to the various organs of the body and are important in the maintenance of blood pressure as their contraction offers resistance to the outflow of blood from the large elastic arteries.

Capillaries receive blood from the arterioles and pass it to the venules. Because their walls are only one cell in thickness oxygen, water and food substances in solution are able to pass from the blood through the walls to supply the tissue cells and waste products can pass back from the tissues to be carried away by the blood.

Capillaries form a network with considerable numbers of them forming anastomoses so that varying amounts of blood may be brought to the part as required.

Venous Return

Blood is collected from the capillary network into venules and then into veins. By the time it reaches the veins it has given up most of its oxygen and is a dark purplish colour. Valves are necessary in the veins to prevent the blood flowing in the wrong direction.

Venous return depends on three factors (Box 16.4).

Maintenance of venous return

- **Suction as the atria relax**
- **Suction during inspiration**—this draws blood towards the heart as well as drawing air into the lungs
- **Pressure on the thin-walled veins by contraction of muscles**—as the lumen of the vein is reduced in size blood would be forced in both directions were it not for the presence of valves which allow the blood to flow towards the heart only

Box 16.4

The force of suction is the most important factor in the maintenance of venous pressure. If the force is lessened, as in heart failure, venous return is impaired and congestion occurs.

Chapter 17
The Circulation

The circulation of the blood is divided into three main parts (Box 17.1).

Divisions of the circulation

- The systemic circulation
- The pulmonary circulation
- The portal circulation

Box 17.1

THE SYSTEMIC CIRCULATION

The vessels which carry the blood from the left ventricle through the body generally to the right atrium are those which constitute the systemic circulation. Arteries may subdivide into several branches at the same point or several branches may be given off in succession. Arteries do not always end in capillaries but may unite with one another, forming *anastomoses*. Examples are found in the brain where two vertebral arteries anastomose to form the basilar artery and two anterior cerebral arteries are connected by the anterior communicating artery (see Fig. 17.2). An enlarged anastomosis may provide a *collateral circulation* if a vessel is occluded by accident or disease; sudden occlusion may be followed by death of the tissue supplied by the vessel whereas gradual occlusion may allow dilation of the anastomosis and adequate nourishment of the tissue. Some arteries have no anastomosis with other arteries and these are called *end-arteries*. Occlusion of an end-artery will cause death (necrosis) of the tissue supplied by the vessel; an example is the central artery of the retina, following occlusion of which permanent blindness will occur.

The Arteries

Arteries are named after the bones in the limbs or from the organ they supply. The most important thing about them is their position;

where they can be compressed against a bone the pulse may be felt or pressure may be applied as a first aid measure in haemorrhage but at the same time it is important to ensure that there is no undue pressure being exerted on them by splints or other equipment. They are usually found on the inner side of a limb where there is only one bone, or between the bones where there are two, and at a joint they would cross the flexor surface. These positions are the safest because the vessels are less likely to be exposed to injury and pressure.

Aorta

The aorta is the main artery of those which carry oxygenated blood to the tissues of the body (Fig. 17.1). It arises from the upper part of

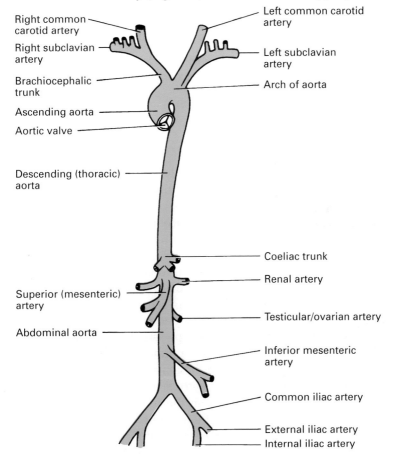

Fig. 17.1 The arch of the aorta and the branches.

the left ventricle, passes upwards and to the right—the *ascending aorta*—and then arches backwards to the left—*the arch of the aorta*—and passes down through the thorax on the left side of the spine—the *descending thoracic aorta*. It enters the abdominal cavity through an opening in the diaphragm called the *aortic hiatus* and is then called the *abdominal aorta*. It ends at the lower border of the fourth lumbar vertebra by dividing into the right and left common iliac arteries.

The *ascending aorta* has two branches, the right and left coronary arteries which arise immediately above the cusps of the aortic valve. They supply the heart wall.

The *arch of the aorta* has three branches which arise from the top of the arch (Box 17.2).

Branches of the arch of the aorta

- **Brachiocephalic trunk**—which divides into the *right subclavian artery* and the *right common carotid artery*
- **Left common carotid artery**
- **Left subclavian artery**

Box 17.2

The *common carotid arteries* supply the head and neck. Each divides into two at the level of the thyroid cartilage to form the external and internal carotid arteries. The *external carotid artery* supplies the outer parts of the face and scalp and has many branches such as the facial, temporal, occipital and maxillary branches. The *internal carotid artery* supplies a large part of the cerebrum, the eyes and the nose and forehead. At the point where the common carotid artery divides there is a dilated area called the *carotid sinus* where a large number of sensory nerve endings, from the glossopharyngeal (9th cranial) nerve, are situated. The sinus reacts to changes in arterial blood pressure and assists in making appropriate changes to return it to normal. A small reddish-brown structure behind the division of the common carotid artery is called the *carotid body*; it acts as a *chemoreceptor* (see p. 217 & 259).

The *vertebral arteries* arise from the first part of the subclavian arteries and pass upwards in the foramina of the transverse processes of the cervical vertebrae; they enter the skull through the foramen magnum (see Fig. 10.2) and join together to form the *basilar artery*.

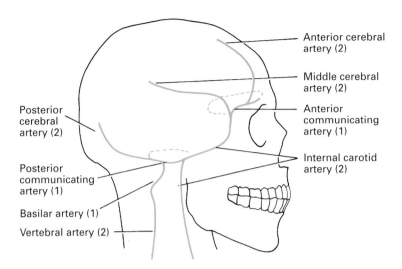

Fig. 17.2 Blood supply to the brain (side view). Broken lines indicate communication with the opposite side of the brain.

An anastomosis named the *circulus arteriosus* connects the vertebral arteries and the two internal carotid arteries. This circle is situated at the base of the brain (Fig. 17.2). In front the two *anterior cerebral arteries* are joined by the *anterior communicating artery* (Fig. 17.3). Behind, the basilar artery, formed by the junction of the two vertebral arteries, divides into two *posterior cerebral arteries*, each of which is joined to the internal carotid arteries by the *posterior communicating artery*. This anastomosis increases the likelihood of an adequate blood supply being maintained to the brain if one of the vessels is injured or occluded.

Blood is supplied to the upper limb through one main artery which is called the *subclavian artery* as far as the outer border of the first rib, the *axillary artery* as far as the middle third of the humerus and the *brachial artery* as far as the back of the radius, where it divides into the radial and ulnar arteries. The *radial artery* passes along the radial side of the forearm to the wrist, where its pulsation can be felt quite easily, and then crosses into the palm of the hand to form the *deep palmar arch* and to unite with the deep branch of the ulnar artery.

The *ulnar artery* runs down the inner side of the forearm to the wrist which it crosses to form the *superficial palmar arch*. The palmar arches supply the digits and each anastomoses freely with the other so that injury to one allows blood to be brought to the part by other vessels.

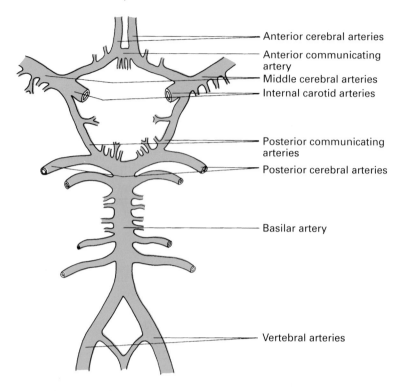

Fig. 17.3   The arterial circle of the cerebrum (circle of Willis).

The *descending thoracic aorta* is situated in the mediastinum; it supplies the pericardium, bronchi, oesophagus, mediastinum, intercostal muscles and breasts through branches which are named after the part they supply.

The *abdominal aorta* begins at the aortic hiatus of the diaphragm, about the level of the last thoracic vertebra. It has many large branch arteries arising from it to supply the abdominal organs and therefore it decreases rapidly in size (Box 17.3 and Fig. 17.4).

Branches of the abdominal aorta

- **Phrenic arteries**—supply the diaphragm
- **Coeliac trunk**—arises just below the aortic opening in the diaphragm and divides into three branches:
  (a) The *left gastric artery*, which supplies the stomach and gives off two or three branches which ascend through the

oesophageal opening in the diaphragm and anastomose with the oesophageal arteries

(b) The *hepatic artery*, which supplies the liver and gives off branches to the duodenum and bile duct as well as supplying the *right gastric artery*

(c) The *splenic artery* which divides into many branches to supply the very vascular spleen and also gives off numerous small vessels to supply the pancreas

- **Superior mesenteric artery**—supplies all the small intestine, except part of the duodenum, and the first part of the large intestine
- **Middle adrenal arteries**—arise, one on each side of the aorta, opposite the superior mesenteric artery and supply the suprarenal glands
- **Renal arteries**—supply the kidneys; the left artery is a little higher than the right because of the positions of the respective kidneys
- **Ovarian arteries** in the female and the **testicular arteries** in the male—supply the organs after which they are named.
- **Inferior mesenteric artery**—supplies the remainder of the large intestine including the sigmoid colon and rectum

Box 17.3

The abdominal aorta divides into two *common iliac arteries* and these again divide, at the level of the last lumbar intervertebral disc, into the *internal iliac artery* which supplies the pelvis, the perineum and the gluteal region and the *external iliac artery* which supplies the lower limb.

Blood is supplied to the lower limb through the *external iliac artery* which crosses the groin and enters the thigh as the *femoral artery*; this in turn becomes the *popliteal artery* at the lower one-third of the thigh. The *popliteal artery* divides into:

1. The *anterior tibial artery*, which runs down the front of the leg on the anterior surface of the interosseous membrane, crosses the front of the ankle joint and supplies the front of the foot as the *dorsalis pedis* artery.

2. The *posterior tibial artery* which runs down the back of the leg to the back of the ankle joint and to the sole of the foot to become the *plantar arch*.

The arteries around the ankle joint anastomose freely with one another to form networks of vessels.

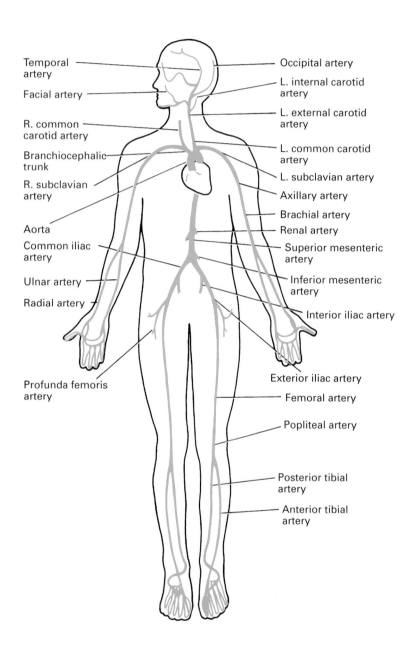

Fig. 17.4 The main arteries.

The Veins

The veins, which return venous blood to the heart, are either superficial veins which vary considerably in position, or deep veins which usually accompany arteries. Arterial pulsation is one of the factors which aids venous return. Systemic veins are more variable than corresponding arteries and anastomoses occur more frequently. In some areas, such as the pelvis and around the vertebral column, the veins form extensive anastomoses and often do not have valves.

The veins can be described in two groups (Box 17.4).

Groups of veins

- The veins of the head, neck, upper limbs and thorax which all end in the superior vena cava
- The veins of the lower limbs, abdomen and pelvis which all end in the inferior vena cava

Box 17.4

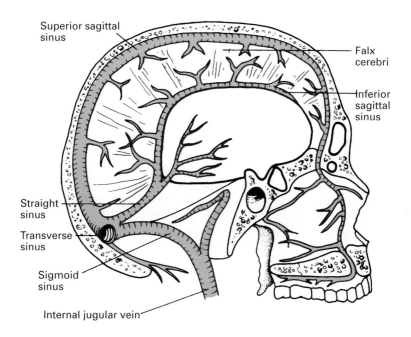

Fig. 17.5 Venous sinuses within the cranium.

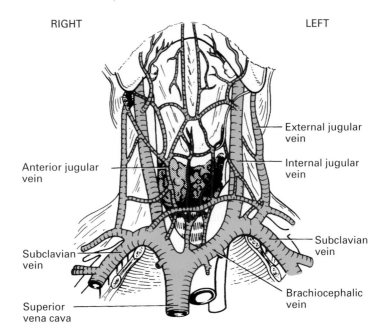

RIGHT

LEFT

External jugular vein

Internal jugular vein

Anterior jugular vein

Subclavian vein

Subclavian vein

Brachiocephalic vein

Superior vena cava

Fig. 17.6 The chief veins of the neck.

The blood from the brain is collected into vessels which are situated between the two layers of the dura mater and are called *venous sinuses*. These in turn empty into the *internal jugular veins* along with blood from the superficial parts of the face and from the neck (Fig. 17.5).

The *external jugular veins* receive blood from the exterior of the cranium and from the deep parts of the face. At the root of the neck the internal jugular veins join with the *subclavian veins* to form the *brachiocephalic veins* which in turn unite to form the superior vena cava through which blood is poured into the right atrium (Fig. 17.6).

The veins of the upper and lower limbs are in two groups: superficial and deep (Box 17.5). The *superficial veins* (Figs 17.7 and 17.8) lie immediately under the skin; the *deep veins* follow the course of the arteries (Fig. 17.9).

Superficial and deep veins of the limbs

Veins of the upper limbs
- *Superficial veins*—the cephalic, basilic and median veins and their tributaries

- *Deep veins*—these pour their blood into the *axillary vein*, which is a continuation of the basilic veins, and which becomes continuous with the subclavian veins

Veins of the lower limbs
- *Superficial veins*—the short and long saphenous veins, which empty into the deep popliteal vein at the back of the knee
- *Deep veins*—the anterior and posterior tibial veins, the popliteal vein and the femoral vein

Box 17.5

Subclavian vein

Axillary vein

Cephalic vein

Basilic vein

Median cephalic vein

Median basilic vein

Cephalic vein

Basilic vein

Median vein

Fig. 17.7 The superficial veins of the forearm.

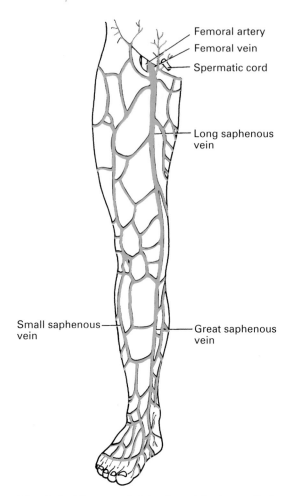

Femoral artery
Femoral vein
Spermatic cord

Long saphenous vein

Small saphenous vein

Great saphenous vein

Fig. 17.8   The superficial veins of the lower limb.

Between the deep and superficial veins of the lower limbs a number of 'perforating' veins exist which have their valves so arranged that normally blood is prevented from flowing from the deep to the superficial veins. If these valves become ineffective blood can flow from the deep to the superficial veins raising the pressure in the latter and causing the dilatation and degeneration known as varicose veins and ulcers.

The femoral vein ends at the inguinal ligament by becoming the *external iliac vein*. The *internal iliac vein*, returning blood from the pelvis, unites with the external iliac vein to become the common iliac vein, one on either side of the body. These unite at the level of

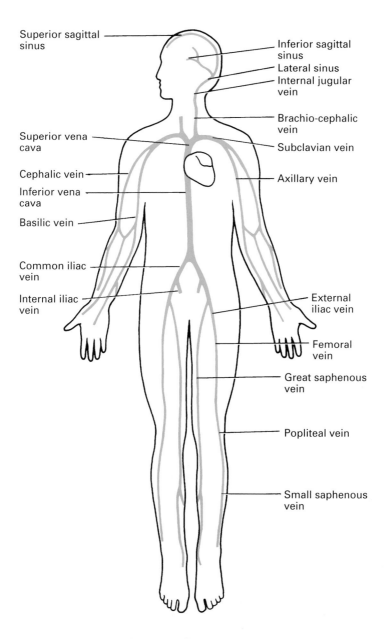

Fig. 17.9 The main veins.

the fifth lumbar vertebra to form the *inferior vena cava* which receives several tributaries, some important ones of which are listed in Box 17.6.

Tributaries of the inferior vena cava

- **Renal veins**—from the kidneys
- **Ovarian or testicular veins**—from the reproductive organs, which empty into the inferior vena cava on the right and into the renal vein on the left
- **Hepatic veins**—return all the blood from the liver, including that from the portal circulation

Box 17.6

The venae cavae pour all venous blood into the right atrium except that from the coronary circulation which is conveyed by the coronary sinus directly into the right atrium.

THE PULMONARY CIRCULATION

These vessels are concerned with carrying deoxygenated blood from the heart to the lungs and oxygenated blood from the lungs back to the heart (see also Chapter 19). The *pulmonary trunk* carries venous blood from the right ventricle; at the left of the fifth thoracic vertebra it divides into *right and left pulmonary arteries* which then branch to carry blood to the various segments of the lungs. These vessels are the only arteries carrying deoxygenated blood.

The four *pulmonary veins* return the oxygenated blood from the lungs to the left atrium, two vessels from each lung. These are the only veins which carry oxygenated blood.

THE PORTAL CIRCULATION

The portal system includes all the veins which drain blood from the abdominal part of the digestive system and from the spleen, pancreas and gall bladder. The blood from these organs is carried to the liver by the *portal vein*, which ends in vessels like capillaries, called *sinusoids*. The blood is then conveyed by the hepatic veins to the inferior vena cava. As mentioned above the liver must also receive oxygenated blood and this is conveyed to it by the hepatic artery.

THE FETAL CIRCULATION

Fetal blood is carried to and from the placenta by umbilical arteries and veins. Most of the blood which enters the right atrium through the inferior vena cava passes through an opening in the atrial septum, called the *foramen ovale*, directly into the left atrium and thence into the left ventricle and aorta. The blood returning to the right atrium through the superior vena cava passes into the right ventricle and pulmonary trunk, after which only a small part of it goes to the lungs. Most of it passes through the *ductus arteriosus* directly to the aorta.

At birth the foramen ovale closes, so that blood cannot pass from the right atrium to the left atrium but is directed into the pulmonary trunk and thence to the lungs. The ductus arteriosus also closes shortly after birth.

Chapter 18
The Lymphatic System

As blood passes through the capillaries in the tissues fluid oozes out through the porous walls and circulates through the tissues themselves, bathing every cell. This fluid is called *tissue* or *interstitial fluid*; it fills the interstices or the spaces between the cells which form the different tissues (Fig. 18.1). It is clear, watery, straw-coloured fluid similar to the plasma of the blood from which it is derived. While blood circulates only through the blood vessels, tissue fluid circulates through the actual tissue and carries nutrients, oxygen and water from the blood stream to each individual cell and carries away its waste products such as carbon dioxide, urea and water, transmitting them to the blood. It is, in other words, the carrying medium between the tissue cells and the blood.

Of the fluid which escapes from the capillaries into the tissues a certain amount passes back through the capillary wall, but its return is more difficult than its escape owing to the constant stream of oncoming blood which fills the capillaries. The excess fluid which cannot return directly into the blood stream is collected up and returned to the blood by a second set of vessels, which form

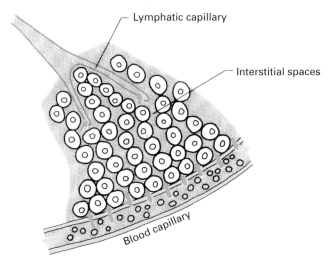

Fig. 18.1 Diagram showing the circulation of tissue fluid. Fluid passes out into the tissues and is collected partly by the blood capillary and partly by the lymph capillary.

the lymphatic system, and the fluid that these vessels contain is called *lymph*.

The lymphatic system comprises four types of structure (Box 18.1).

Components of the lymphatic system

- Lymphatic capillaries
- Lymphatic vessels
- Lymphatic nodes
- Lymphatic ducts

Box 18.1

Lymphatic Capillaries

The lymphatic capillaries arise in the spaces in the tissues as fine hair-like vessels with porous walls. These gather up excess fluid from the tissues and unite to form the lymphatic vessels. The walls of lymph capillaries are permeable to substances of greater molecular size than those which can pass through the walls of blood capillaries.

Lymphatic Vessels

The lymphatic vessels are thin-walled, collapsible tubes similar in structure to the veins, but carrying lymph instead of blood. They are finer and more numerous than the veins and, like them are provided with valves to prevent the lymph moving in the wrong direction (Fig. 18.2). Lymphatic vessels are found in most tissues except the central nervous system, but run particularly in the subcutaneous tissues and pass through one or more lymphatic nodes (Figs 18.3–18.5).

Lymphatic Nodes

The lymphatic nodes are small bodies varying in size from a pin-head to an almond. Lymphatic vessels bring lymph to them and are

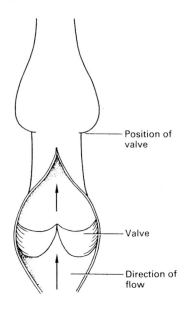

Fig. 18.2   A lymphatic vessel (cut open).

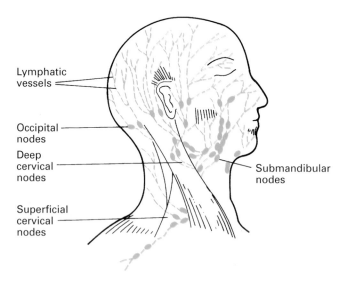

Fig. 18.3   The lymphatic vessels of the head and neck.

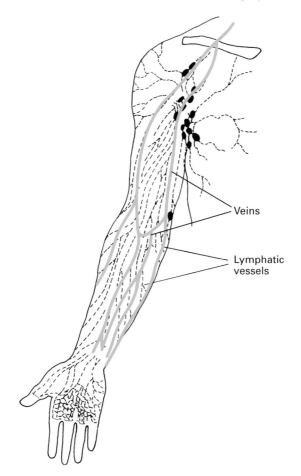

Veins

Lymphatic
vessels

Fig. 18.4 The lymphatic vessels of the upper limb.

called afferent vessels. These enter and divide up within the node and discharge the lymph into the lumen. The lymph is then gathered up again into fresh lymphatic vessels called efferent vessels, which carry it on and ultimately empty it into the lymphatic ducts after possibly carrying the lymph through more nodes (Fig. 18.6). The lymphatic nodes consist essentially of cells similar to the white blood corpuscles (lymphocytes), held together by a network of connective tissue, which also forms a capsule for the node.

The functions of the lymphatic nodes are listed in Box 18.2.

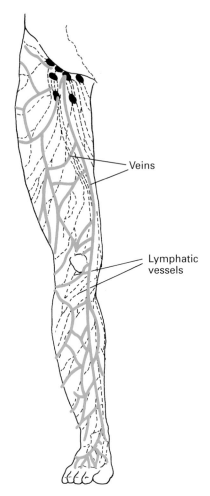

Fig. 18.5   The lymphatic vessels of the lower limb.

The functions of the lymphatic nodes

- **To filter the lymph of bacteria as it passes through.** Thus when tissues are infected the node may become swollen and tender. If the infection is a mild one, the organisms will be overcome by cells of the node and the tenderness and swelling will subside. If it is severe the organisms will cause acute inflammation and destruction of the white cells may result, causing an abscess to form in the node. If the bacteria

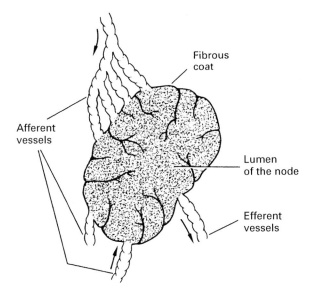

Fig. 18.6 A lymphatic node, showing the afferent and efferent vessels.

are not destroyed by the node they may pass on in the lymph stream and infect the general circulation, causing septicaemia
- **To provide fresh lymphocytes for the blood stream.** The cells of the node constantly multiply and the newly formed cells are carried away in the lymph.
- **They may produce some antibodies and antitoxins** to prevent infection

Box 18.2

The lymphatic nodes are for the most part massed together in groups in various parts of the body. Groups in the neck and under the chin filter the lymph from the head, tongue and the floor of the mouth. A group in the axilla filters the lymph from the upper limb and the chest wall. A group in the groin filters lymph from the lower limb and lower abdominal wall. Groups within the thorax and abdomen filter the lymph from the internal organs.

Special areas where much lymphatic tissue is found include: the palatine and pharyngeal tonsils; the thymus gland; aggregated lymphatic follicles in the small intestine; the appendix and the spleen.

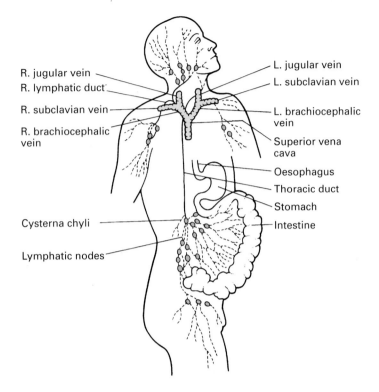

R. jugular vein
R. lymphatic duct
R. subclavian vein
R. brachiocephalic vein
Cysterna chyli
Lymphatic nodes

L. jugular vein
L. subclavian vein
L. brachiocephalic vein
Superior vena cava
Oesophagus
Thoracic duct
Stomach
Intestine

Fig. 18.7   The lymphatic ducts.

Lymphatic Ducts

After filtration by the nodes the lymph is emptied by the lymphatic vessels into the two lymphatic ducts: the thoracic duct and right lymphatic duct (Fig. 18.7).

The *thoracic duct* is the larger. It begins in a small pouch at the back of the abdomen called the *cisterna chyli*. Into this empty all the lymphatic vessels from the lower limbs and the abdominal and pelvic organs. From the cisterna chyli the duct runs up through the mediastinum behind the heart to the root of the neck, where it turns to the left, is joined by the lymphatic vessels from the left side of the head and thorax and the left upper limb, and finally empties into the *left* subclavian vein at its junction with the left internal jugular vein. It is about 45 cm long and is provided with valves to prevent the lymph flowing in the wrong direction.

The *right lympahtic duct* is a comparatively small vessel formed by the joining of the lymphatic vessels from the right side of the head and thorax and the right upper limb at the root of the neck. It is only about 1 cm long and enters into the *right subclavian vein*, where it joins the right internal jugular vein.

The lymphatic ducts thus gather up all the lymph and return it to the blood stream, from which the fluid in the tissues is constantly renewed.

THE FUNCTIONS OF THE LYMPHATIC SYSTEM

The functions of the lymphatic system are listed in Box 18.3.

Functions of the lymphatic system

- The lymphatic vessels gather up the excessive fluid or lymph from the tissues and thus permit a constant stream of fresh fluid to circulate through them
- It is the channel by which excess proteins in the tissue fluid pass back into the blood stream
- The nodes filter the lymph of bacterial infection and harmful substances
- The nodes produce fresh lymphocytes for the circulation
- The lymphatic vessels in the abdominal organs assist in the absorption of digested food, especially fat

Box 18.3

THE MECHANISM OF THE LYMPHATIC CIRCULATION

The lymphatic circulation is maintained partly by suction, partly by pressure. Suction is the more important factor. The lymphatics empty into the large veins approaching the heart, and here there is negative pressure due to suction as the heart expands, and also suction towards the thorax during the act of inspiration.

Pressure is exerted on the lymphatics, as on the veins, by the contraction of muscles, and this outside pressure drives lymph onwards because the valves prevent a backward flow. There is also a slight pressure from the fluid in the tissues, on account of the constant pouring out of fresh fluid from the capillaries. If there is obstruction to the flow of the lymph through the lymphatic

Fig. 18.8   The spleen and its blood vessels.

system, oedema results, i.e. a swelling of the tissues due to excess fluid collecting in them. The condition also results from any obstruction of the veins, since these also drain fluid from the tissues.

THE SPLEEN

The spleen is a large nodule of lymphoid tissue (Fig. 18.8). By function it belongs to the circulatory system, as do the lymphatic nodes. It is a deep purplish-red colour, and lies high up at the back of the abdomen on the left side behind the stomach. It is enclosed in a capsule of fibrous tissue, and fibrous strands make a supporting meshwork throughout the gland. The spaces of this meshwork are filled in with a pulp-like material, called the splenic pulp, which is the essential substance of the organ and contains cells of different types. Many of these are similar to the lymphocytes of the blood and lymph nodes, and these help to produce fresh white cells for the blood stream. Others are phagocytes or devouring cells which engulf the red blood cells that are beginning to wear out and break them down.

The *functions of the spleen* are not fully known, but it is thought to be:

1.   A source of fresh lymphocytes for the blood stream
2.   A seat for the destruction of red blood cells

The spleen is also thought to assist in fighting infection, as it becomes enlarged in certain diseases where the blood is infected, e.g. malaria and typhoid fever. It probably helps in the manufacture of antibodies to fight infection. It is not essential to life, and can be removed by operation when it is responsible for ill health, e.g. in haemolytic anaemia.

Chapter 19
The Respiratory System

All living cells require a constant supply of oxygen in order to carry on their metabolism. Oxygen is in the air, and the respiratory system is constructed in such a way that air can be taken into the lungs, where some of the oxygen is extracted for use by the body and at the same time carbon dioxide and water vapour are given up. The organs of the respiratory system are listed in Box 19.1 and shown in Fig. 19.1.

Organs of the respiratory tract

- The nose ⎫
- The pharynx ⎬ leading to the lungs
- The larynx ⎪
- The trachea ⎭
- The bronchi ⎫
- The bronchioles ⎬ within the lungs
- The alveolar ducts and alveoli ⎭

Box 19.1

THE NOSE

The external nose is the visible part of the nose, formed by the two nasal bones and by cartilage. It is both covered and lined by skin and inside there are hairs which help to prevent foreign material from entering. The nasal cavity (Fig. 19.2) is a large cavity divided by a septum. The anterior nares are the openings which lead in from without and the posterior nares are similar openings at the back, leading in to the pharynx. The roof is formed by the ethmoid bone at the base of the skull and the floor by the hard and soft palates at the roof of the mouth. The lateral walls of the cavity are formed by the maxilla, the superior and middle nasal conchae of the ethmoid bone and the inferior nasal concha. The posterior part of the dividing septum is formed by the perpendicular plate of the

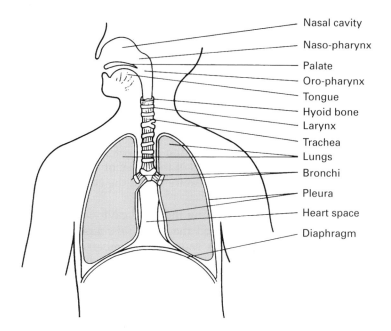

Fig. 19.1 Diagram of the respiratory tract.

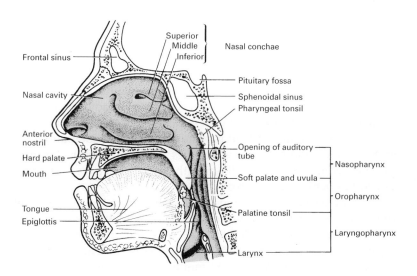

Fig. 19.2 Saggital section of the nose, mouth, pharynx and larynx.

ethmoid bone and by the vomer, while the anterior part is made of cartilage.

The three nasal conchae project into the nasal cavity on each side and greatly increase the surface area of the inside of the nose. The cavity of the nose is lined throughout with ciliated mucous membrane, which is extremely vascular, and atmospheric air is warmed as it passes over the epithelium, which contains many capillaries. The mucus moistens the air and entraps some of the dust and the cilia move the mucus back into the pharynx for swallowing or expectoration. The nerve endings of the sense of smell are situated in the highest part of the nasal cavity round the cribriform plate of the ethmoid bone.

Some of the bones surrounding the nasal cavity are hollow. The hollows in the bones are called the *paranasal sinuses*, which both lighten the bones and act as sounding chambers for the voice, making it resonant. The maxillary sinus lies below the orbit and opens through the lateral wall of the nose. The frontal sinus lies above the orbit towards the midline of the frontal bone. The ethmoidal sinuses are numerous and are contained within the part of the ethmoid bone separating the orbit from the nose, and the sphenoidal sinus is in the body of the sphenoid bone. All the paranasal sinuses are lined with mucous membrane and all open into the nasal cavity, from which they may become infected.

THE PHARYNX

The roof of the pharynx is formed by the body of the sphenoid bone and inferiorly it is continuous with the oesophagus (Fig. 19.2). At the back it is separated from the cervical vertebrae by loose connective tissue, while the front wall is incomplete and communicates with the nose, mouth and larynx. The pharynx is divided into three sections, the naso-pharynx which lies behind the nose, the oro-pharynx which lies behind the mouth, and the laryngeal pharynx which lies behind the larynx.

The *naso-pharynx* is that part of the pharynx which lies behind the nose above the level of the soft palate. On the posterior wall there are patches of lymphoid tissue called the pharyngeal tonsils, commonly referred to as adenoids. This tissue sometimes enlarges and blocks the pharynx causing mouth breathing in children. The auditory tubes open from the lateral walls of the naso-pharynx and through them air is carried to the middle ear. The naso-pharynx is lined with ciliated mucous membrane which is continuous with the lining of the nose.

The *oro-pharynx* lies behind the mouth below the level of the soft palate, with which its lateral walls are continuous. Between the folds of these walls, which are called the palato-glossal arches, are collections of lymphoid tissue called the palatine tonsils. The oro-pharynx is part of both the respiratory tract and the alimentary tract, but it cannot be used for swallowing and breathing simultaneously. During the act of swallowing breathing stops momentarily and the oro-pharynx is completely blocked off from the naso-pharynx by the raising of the soft palate. The oro-pharynx is lined with stratified epithelium.

THE LARYNX

The larynx is continuous with the oro-pharynx above and with the trachea below (Fig. 19.2). Above it lies the hyoid bone and the root of the tongue. The muscles of the neck lie in front of the larynx, and

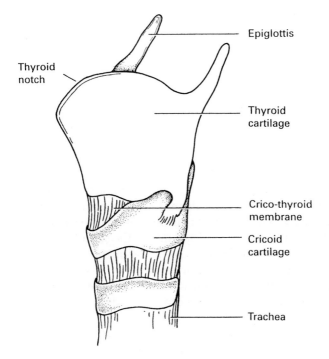

Epiglottis

Thyroid notch

Thyroid cartilage

Crico-thyroid membrane

Cricoid cartilage

Trachea

Fig. 19.3   The laryngeal cartilages.

behind the larynx lies the laryngo-pharynx and the cervical verte-brae. On either side are the lobes of the thyroid gland. The larynx is composed of several irregular cartilages joined together by ligaments and membranes (Fig. 19.3).

The *thyroid cartilage* is formed of two flat pieces of cartilage fused together in the front to form the laryngeal prominence or Adam's apple. Above the prominence is a notch called the thyroid notch. The thyroid cartilage is larger in the male than in the female. The upper part is lined with stratified epithelium and the lower part with ciliated epithelium.

The *cricoid cartilage* lies below the thyroid cartilage and is shaped like a signet ring with the broad portion at the back. It forms the lateral and posterior walls of the larynx and is lined with ciliated epithelium.

The *epiglottis* is a leaf-shaped cartilage attached to the inside of the front wall of the thyroid cartilage immediately below the thyroid notch. During swallowing the larynx moves upwards and forwards so that its opening is occluded by the epiglottis.

The *arytenoid cartilages* are a pair of small pyramids made of hyaline cartilage. They are situated on top of the broad part of the cricoid cartilage, and the vocal ligaments are attached to them. They form the posterior wall of the larynx.

The hyoid bone and the laryngeal cartilages are joined together by ligaments and membranes. One of these, the *cricothyroid membrane*, is attached all round to the upper edge of the cricoid cartilage and has a free upper border, which is not circular like the lower border, but makes two parallel lines running from front to back. The two parallel edges are the *vocal ligaments* (Fig. 19.4). They are fixed to the middle of the thyroid cartilage in front and to the arytenoid cartilages behind, and they contain much elastic tissue. When the intrinsic muscles of the larynx alter the position of the arytenoid cartilages the vocal ligaments are pulled together, nar-rowing the gap between them. If air is forced through the narrow gap, called the chink, during expiration, the vocal ligaments vibrate and sound is produced. The pitch of the sound produced depends on the length and tightness of the ligaments; an increased tension gives a higher note, a slacker tension a lower note. Loudness depends on the force with which the air is expired. The alteration of the sound into different words depends on the movements of the mouth, tongue, lips and facial muscles.

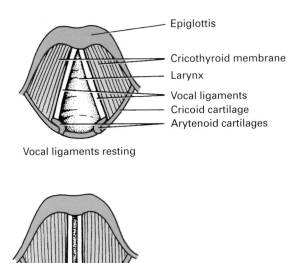

Epiglottis

Cricothyroid membrane

Larynx

Vocal ligaments
Cricoid cartilage
Arytenoid cartilages

Vocal ligaments resting

Vocal ligaments during speech

Fig. 19.4 Diagram of the vocal ligaments.

THE TRACHEA

The trachea begins below the larynx and runs down the front of the neck into the chest. It divides into the right and the left main bronchi at the level of the fifth thoracic vertebra. It is about 12 cm long. The isthmus of the thyroid gland crosses in front of the upper part of the trachea, and the arch of the aorta lies in front of the lower part, with the manubrium of the sternum in front of it. The oesophagus lies behind the trachea, separating it from the bodies of the thoracic vertebrae. On either side of the trachea lie the lungs, with the lobes of the thyroid gland above them. The wall of the trachea is made of involuntary muscle and fibrous tissue strengthened by incomplete rings of hyaline cartilage. The deficiency in the cartilage lies at the back where the trachea is in contact with the oesophagus. When a bolus of food is swallowed the oesophagus is able to expand without hindrance, but the cartilage maintains the patency of the airway. The trachea is lined with ciliated epithelium containing goblet cells which secrete mucus. The cilia sweep the mucus and foreign particles upwards towards the larynx.

THE LUNGS

The lungs are two large spongy organs lying in the thorax on either side of the heart and great vessels. They extend from the root of the neck to the diaphragm and are roughly cone-shaped with the apex above and the base below. The ribs, costal cartilages and intercostal muscles lie in front of the lungs and behind them are the ribs, the intercostal muscles and the transverse processes of the thoracic vertebrae. Between the lungs is the *mediastinum*, which completely separates one side of the thoracic cavity from the other, stretching from the vertebrae behind to the sternum in front. Within the mediastinum lie the heart and great vessels, the trachea and the oesophagus, the thoracic duct and the thymus gland. The lungs are divided into lobes (Fig. 19.5). The left lung has two lobes, separated by the oblique fissure. The superior lobe is above and in front of the inferior lobe which is conical in shape. The right lung has three lobes. The inferior lobe is separated by an oblique fissure in a similar manner to the left inferior lobe. The remainder of the lung is separated by a horizontal fissure into the superior lobe and the middle lobe. Each lobe is further divided into named

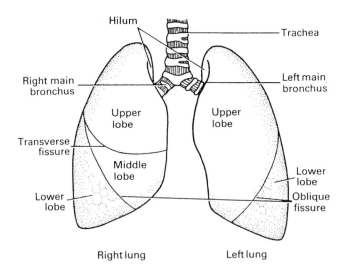

Fig. 19.5 Diagram showing the lobes of the lungs.

broncho-pulmonary segments, separated from each other by a wall of connective tissue and each having an artery and a vein. Each segment is also divided into smaller units called lobules.

The Bronchi

The two main bronchi commence at the bifurcation of the trachea and one leads into each lung. The left main bronchus is narrower, longer and more horizontal than the right main bronchus because the heart lies a little to the left of the midline. Each main bronchus divides into branches, one for each lobe. Each of these then divides into named branches, one for each broncho-pulmonary segment, and divides again into progressively smaller bronchi within the lung substance. The bronchi are similar in structure to the trachea but the cartilage is less regular.

The Bronchioles

The finest bronchi are called bronchioles. They have no cartilage but are composed of muscular, fibrous and elastic tissue lined with cuboid epithelium. As the bronchioles become smaller, the muscular and fibrous tissue disappears and the smallest tubes, called terminal bronchioles, are a single layer of flattened epithelial cells.

The Alveolar Ducts and Alveoli

The terminal bronchioles branch repeatedly to form minute passages called alveolar ducts, from which alveolar sacs and alveoli open. The alveoli are surrounded by a network of capillaries (Fig.

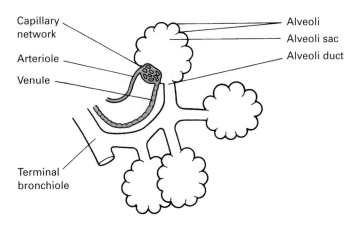

Fig. 19.6 Diagram of alveoli.

19.6). Deoxygenated blood enters the capillary network from the pulmonary artery and oxygenated blood leaves it to enter the pulmonary veins. It is in the capillary network that the exchange of gases takes place between the air in the alveoli and the blood in the vessels.

The Hilum of the Lung

The hilum is a triangular-shaped depression on the concave medial surface of the lung. The structures forming the root of the lung enter and leave at the hilum, which is at the level of the fifth to seventh thoracic vertebrae. These structures include the main bronchus, the pulmonary artery, the bronchial artery and branches of the vagus nerve, which enter at this point, and two pulmonary veins, the bronchial veins and lymphatic vessels, which leave the lung at the root. There are also many lymph nodes round the root of the lung.

The Pleura

The pleura is a serous membrane which surrounds each lung. It is composed of flattened epithelial cells on a basement membrane and it has two layers (Fig. 19.7). The visceral pleura is firmly attached to the lungs, covering their surfaces and dipping into the inter-lobar fissures. At the root of the lungs the visceral layer

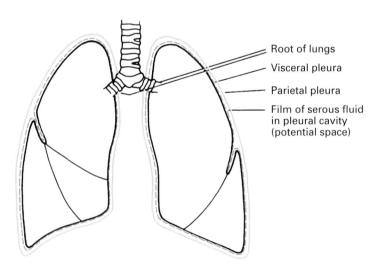

Root of lungs
Visceral pleura
Parietal pleura
Film of serous fluid in pleural cavity (potential space)

Fig. 19.7   Diagram of the pleura.

is reflected back to become the parietal layer which lines the chest wall and covers the superior surface of the diaphragm. The two layers of the pleura are normally in close contact with each other, separated only by a film of serous fluid which enables them to glide over one another without friction. This potential space between the layers is called the pleural cavity.

MECHANISM OF RESPIRATION

Respiration consists of two parts, inspiration and expiration. The chest expands during inspiration owing to movement of the diaphragm and the intercostal muscles. When the diaphragm contracts during inspiration it is flattened and lowered and the thoracic cavity is increased in length (see Chapter 14). The external intercostal muscles, on contraction, lift the ribs and draw them out, increasing the depth of the thoracic cavity. As the chest wall moves up and out the parietal pleura, which is closely attached to it, moves with it. The visceral pleura follows the parietal pleura and the volume of the interior of the thorax is increased. The lung expands to fill the space and air is sucked into the bronchial tree.

Expiration during quiet breathing is passive. The diaphragm relaxes and assumes its original domed shape. The intercostal muscles relax and the ribs revert to their previous position. The lungs recoil and air is driven out through the bronchial tree. In forced expiration the internal intercostal muscles contract actively to lower the ribs. The accessory muscles of respiration may be brought into use during deep breathing or when the airway is obstructed. During inspiration the sternocleidomastoid muscles raise the sternum and increase the diameter of the thorax from front to back. Serratus anterior and pectoralis major pull the ribs outwards when the arm is fixed. Latissimus dorsi and the muscles of the anterior abdominal wall help to compress the thorax during forced expiration.

Control of Respiration

Respiration is controlled by the respiratory centre in the medulla oblongata. The presence of an accumulation of carbon dioxide in the blood stimulates chemoreceptors in the great arteries. Impulses are carried by the vagus and glossopharyngeal nerves to the respiratory centre, from the respiratory centre by the phrenic nerves to the diaphragm and by the intercostal nerves to the

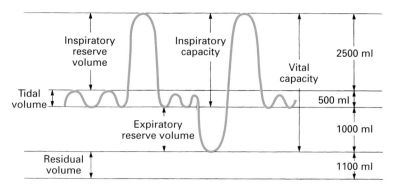

Fig. 19.8   Lung volumes and capacities.

intercostal muscles. These impulses cause the muscles to contract and inspiration occurs.

THE CAPACITY OF THE LUNGS

Tidal volume is the amount of air breathed in and out during normal quiet breathing (about 500 ml). After a normal expiration the amount of air taken in during forced inspiration is called the *inspiratory capacity* (about 3000 ml). It includes the tidal volume. After a quiet expiration it is possible to force another 1000 ml of air from the lungs. This is the *expiratory reserve volume* (Fig. 19.8).

After the deepest possible expiration there is still the *residual volume* (about 1100 ml) left in the respiratory passages.

Vital capacity is the largest volume of air expired after the deepest possible inspiration (about 4000 ml). An important measure of respiratory function is the amount of forced expiration in one second. It is called the *forced expiratory volume* and should measure 75 to 80 per cent of the vital capacity.

GAS EXCHANGE

Exchange of gases within the body takes place both in the lungs, where it is called *external respiration*, and in the tissues, where it is called *internal respiration*. An elementary law of physics relating to gases states that they tend to diffuse from a higher pressure to a lower pressure (Fig. 19.9). The air which is breathed in contains several gases. The composition of inspired air is shown in Box 19.2.

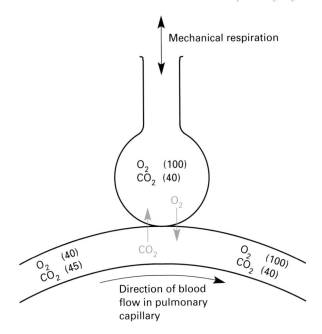

Fig. 19.9 Gas exchange in the lungs. (All values in parentheses are in units of mmHg.)

The composition of inspired air

- Nitrogen   79 per cent
- Oxygen   21 per cent
- Carbon dioxide   0·04 per cent
- Water vapour
- Traces of other gases

Box 19.2

External Respiration

When inspired air reaches the alveoli it is in close contact with the blood in the surrounding capillary network. Oxygen in the alveoli, at a pressure of 100 mmHg, comes in contact with oxygen in the venous blood at a pressure of 40 mmHg and therefore the gas diffuses into the blood until the pressures are equal. At the same time carbon dioxide in the blood, at a pressure of 46 mmHg, comes in contact with alveolar carbon dioxide at a pressure of 40 mmHg and therefore the gas diffuses out of the blood into the alveoli. The gaseous content of expired air is therefore altered to contain less oxygen and more carbon dioxide. The nitrogen content remains the same. The composition of expired air is given in Box 19.3.

The composition of expired air

- Nitrogen   79 per cent
- Oxygen   16 per cent
- Carbon dioxide   4·5 per cent
- Water vapour
- Traces of other gases

Box 19.3

Internal Respiration

The oxygen which has diffused into the blood is carried in the haemoglobin—now called oxyhaemoglobin—to the tissues. Here the pressure of the oxygen is low, so the gas diffuses out of the blood and into the tissues, the amount depending on the activity in the tissue. At the same time carbon dioxide, produced in the tissues, is carried away by the blood.

Oxygen–Haemoglobin Dissociation

The ability of haemoglobin to bind oxygen at the alveoli and then to release it into the peripheral tissues in a way which is physiologically useful is demonstrated by the *oxygen–haemoglobin dissociation curve* (Fig. 19.10). This curve is obtained by measuring the amount of haemoglobin which has bound oxygen (the *saturation with oxygen*) as the amount of oxygen (the *partial pressure of oxygen*) in the blood is increased. The curve has a characteristic sigmoidal (S) shape. If the curve were hyperbolic it would mean that oxygen was not released in the peripheral tissues until the levels of oxygen has fallen to physiologically dangerous ones. The sigmoidal shape means that oxygen is readily released into peripheral tissues from the haemoglobin in response to falling oxygen levels in order to maintain levels of oxygen which allow these tissues to function. The curve is sensitive to conditions in the

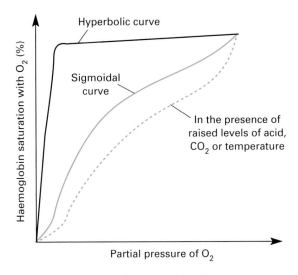

Fig. 19.10 Oxygen–haemoglobin dissociation curve.

peripheral tissues; it moves to the right in response to increased levels of acidity, carbon dioxide or temperature, all of which indicate that there is increased metabolic activity and, therefore, a greater demand for oxygen.

SECTION V: NUTRITION AND ELIMINATION

Section Introduction

Most of this section is concerned with the digestive tract and its associated organs. The digestive tract runs from the mouth to the rectum and is responsible for the ingestion, mixing and mechanical breakdown and absorption of nutrients. Ultimately, those contents of our food which are not absorbed are expelled at the rectum.

The organs associated with digestion are the liver, the pancreas and the gall bladder and these produce and store secretions which are released into the digestive tract and which facilitate the mechanical and chemical breakdown of food. The latter process is called 'digestion' and this prepares food for absorption and subsequent involvement in metabolism. The main features of metabolism are considered.

While it is not anatomically linked to the digestive system, the main function of the urinary system is relevant here as it is responsible for ridding the body of excess toxic substances. Principally these are expelled in urine in the form of urea. The urinary system also responds to changing levels of fluid in the body and can, thereby, conserve water or expel excess water.

Chapter 20
The Digestive System

The digestive system (Fig. 20.1) consists of all the organs which are concerned in the chewing, swallowing, digestion and absorption of food and in the elimination from the body of indigestible and undigested food.

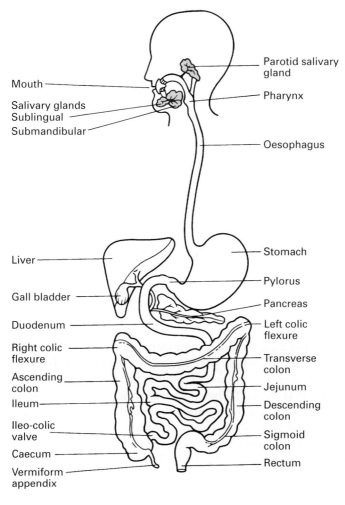

Fig. 20.1 Diagrammatic representation of the digestive system.

It consists of the digestive tube or alimentary canal and the accessory organs of digestion.

The *digestive tract* is about nine metres long and consists of six parts (Box 20.1). The *accessory organs* of digestion are listed in Box 20.2.

The digestive tract

- The mouth
- The pharynx
- The oesophagus
- The stomach
- The small intestine
- The large intestine, which reaches the surface of the body at the anus

Box 20.1

Accessory organs of digestion

- The teeth
- Three pairs of salivary glands
- The liver and bile ducts (see Chapter 21)
- The pancreas (see Chapter 21)

Box 20.2

THE MOUTH

The mouth is a cavity bounded externally by the lips and cheeks and leading into the pharynx. The roof is formed by the hard and soft palates and the anterior two-thirds of the tongue fills the floor of the mouth (Fig. 20.2). The walls are formed by the muscles of the cheeks. The mucous membrane which lines the mouth is continuous with the skin of the lips and with the mucous lining of the pharynx. The lips enclose the orbicularis oris muscle which keeps the mouth closed. The *hard palate* is formed by parts of the palatine bones and the maxillae; its upper surface forms the floor of the nasal cavity. The *soft palate* is suspended from the posterior border of the hard palate and extends down between the oral and nasal

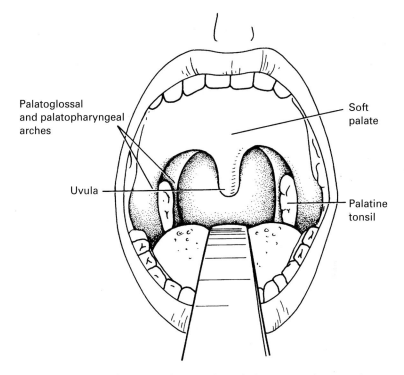

Fig. 20.2 Looking into the mouth with the tongue depressed.

parts of the pharynx. Its lower border hangs like a curtain between the mouth and the pharynx and a small conical process, called the *uvula*, hangs down from it. Two curved folds of mucous membrane extend sideways and downwards from each side of the base of the uvula, called the *palatoglossal and palatopharyngeal arches*, between which lie the masses of lymphoid tissue known as the *palatine tonsils*.

The Tongue

The tongue is a muscular organ which is attached to the hyoid bone and the mandible. It is covered in certain areas with modifications of the mucous membrane which appear as projections to increase the surface area and are called *papillae*. In addition specialized areas called *taste buds* (Fig. 20.3) are widespread over almost the entire area of the tongue. The under surface of the anterior part of the tongue is connected to the floor of the mouth by a fold of mucous membrane called the *frenulum*. The functions of the tongue are:

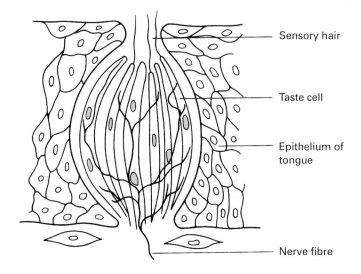

Sensory hair

Taste cell

Epithelium of tongue

Nerve fibre

Fig. 20.3 A taste bud in the tongue.

1. It is the organ of *taste*
2. It assists in the *mastication* of food
3. It assists in *swallowing*
4. It assists with *speech*

The Teeth

Man is provided with two sets of *teeth* which make their appearance at different periods of life. The first set are *deciduous or primary teeth* and erupt through the gums during the first and second years of life. The second set begin to replace the first about the sixth year and the process is usually complete by the twenty-fifth year. Since they cannot be replaced, and may be retained until old age, they are known as the *permanent teeth*.

Each tooth consists of three parts: (Box 20.3 and Fig. 20.4).

The parts of a tooth

- **Crown**—projecting beyond the gum
- **Root**—embedded in the alveolus of the maxilla or mandible (a tooth may have one, two or three roots)
- **Neck**—the constricted part between the crown and the root

Box 20.3

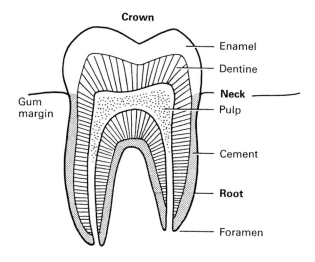

Fig. 20.4   Section of a tooth, showing its structure.

In the centre of all these parts is the *pulp* and immediately outside the pulp is a yellowish-white layer, called *dentine*, which forms the main part of the tooth. The outer layer of the tooth is in two parts; that covering the crown is called *enamel* and is a hard, white layer while that covering the root is called *cement* and is a thin layer resembling bone in structure. The pulp is richly supplied with blood vessels and nerves which enter the tooth through foramina at the apex of each root.

There are four types of teeth (Box 20.4).

Types of teeth

- **Incisor teeth** have chisel-shaped crowns giving a sharp cutting edge for biting food
- **Canine teeth** have large conical crowns
- **Premolar (or bicuspid) teeth** have almost circular crowns with two cusps for grinding food
- **Molar teeth** are the largest, and they have broad crowns with four or five cusps

Box 20.4

There are 20 deciduous teeth and 32 permanent teeth (Table 20.1).

Table 20.1 The Deciduous Teeth

	Molars	Canines	Incisors	Canines	Molars
Upper jaw	2	1	2 2	1	2
Lower jaw	2	1	2 2	1	2

The Permanent Teeth

	Molars	Premolars	Canines	Incisors	Canines	Premolars	Molars
Upper jaw	3	2	1	2 2	1	2	3
Lower jaw	3	2	1	2 2	1	2	3

The Salivary Glands

There are three pairs of *salivary glands*. The *parotid gland* is the largest and lies just below the ear; its duct is about 5 cm long and opens into the mouth opposite the second upper molar tooth. It is this gland which is affected by the disease commonly known as mumps. The *submandibular gland* and the *sublingual gland* both open into the floor of the mouth. *Saliva* is secreted reflexly by the presence of food in the mouth and by a conditioned (or learned) reflex which enables saliva to be secreted by the sight, smell or thought of food. Saliva contains a large amount of water which moistens and softens the food; mucus, which combines the food and lubricates it for its passage down the oesophagus; and the enzyme *salivary amylase*, which acts on cooked starch (carbohydrate) and splits it into maltose and dextrin. Saliva also cleanses the mouth and teeth and keeps the soft parts supple.

THE PHARYNX

When the food is well chewed and moistened the tongue rolls it into a *bolus* and carries it towards the *oral part of the pharynx*. The soft palate rises up to occlude the naso-pharynx and the epiglottis moves upwards and forwards so that the bolus passes over the closed inlet of the larynx and on into the *largyngeal part of the pharynx* and thence to the oesophagus. This is an excellent example of muscular coordination and if it is not achieved correctly 'choking' will result (see Fig. 19.2).

THE OESOPHAGUS

The oesophagus is a muscular canal about 25 cm long extending from the pharynx to the stomach. It begins at the level of the sixth cervical vertebra and descends through the mediastinum in front of the vertebral column and behind the trachea. It passes through the diaphragm at the level of the tenth thoracic vertebra and ends at the cardiac orifice of the stomach at the level of the eleventh thoracic vertebra. On each side of the upper part of the oesophagus are the corresponding common carotid artery and part of the thyroid gland. The oesophagus has four layers and is similar in structure to the remainder of the alimentary canal (Box 20.5 and Fig. 20.5).

Layers of the alimentary canal

- **Fibrous outer layer**—consists of areolar tissue containing many elastic fibres
- **Muscular layer**—has two layers, the outer fibres running longitudinally and the inner layer consisting of circular fibres
- **Areolar or submucous layer**—connects the mucous and muscular coats and contains the larger blood vessels and nerves, as well as the mucous glands
- Inner lining of **mucous membrane** secretes mucus

Box 20.5

The muscular coat of the upper two-thirds of the oesophagus is of striped voluntary muscle; the lower one-third contains unstriped involuntary muscle. The oesophagus is innervated by the vagus nerve. Movement of food through the oesophagus is by peristaltic action. *Peristalsis* means a wave of dilatation followed by a wave of contraction as succeeding muscle fibres relax and contract. It takes about nine seconds for a wave of peristalsis to pass the bolus of food from the pharynx to the stomach.

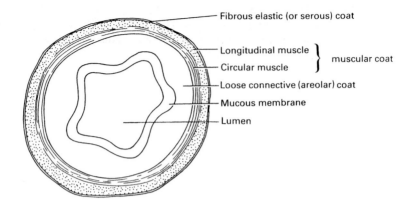

Fig. 20.5 Cross-section of the alimentary tract.

THE STOMACH

The stomach is the most dilated part of the digestive tube and is situated between the end of the oesophagus and the beginning of the small intestine. Its shape and position are altered by changes within the abdominal cavity and by the stomach contents but it lies below the diaphragm, slightly to the left of the midline (Figs 20.6 and 20.7).

The stomach is approximately J-shaped and has two curvatures. The *lesser curvature* forms the right (or posterior) border of the

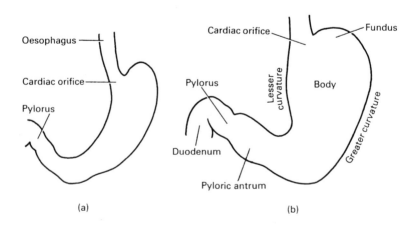

Fig. 20.6 The stomach: (a) empty; (b) containing food,

Fig. 20.7 The stomach. An X-ray of a barium meal, showing the normal stomach, the constriction of the pylorus and the 'duodenal cap', i.e. the beginning of the duodenum, filled with barium. (1) Body of stomach; (2) pylorus; (3) duodenal cap.

stomach. The *greater curvature* is directed mainly forwards and first forms an arch upwards and to the left to form the *fundus* of the stomach; it then passes downwards and finally turns right to the point where it joins the duodenum. The capacity of the stomach is about 1500 ml in the adult.

The upper opening from the oesophagus is called the *cardiac orifice* and the circular muscle fibres of the oesophagus are slightly thicker at this point and consistute a weak sphincter muscle.

The lower opening, into the duodenum, is called the *pyloric orifice* and it is guarded by the strong *pyloric sphincter* which prevents regurgitation of food from the duodenum into the stomach.

The wall of the stomach consists of four layers (Box 20.6 and Fig. 20.8).

The layers of the stomach

- The outer **serous layer**—the visceral layer of the peritoneum (see p. 291)
- The **muscular layer**—this consists of three layers of smooth muscle fibres, the outer being longitudinal, the middle being circular and the inner being oblique
- The **submucous layer**—this is composed of loose areolar tissue
- The lining of **mucous membrane**—this is honeycombed in appearance because of the presence of the gastric glands and their openings. The mucous membrane has numerous folds, called *rugae*, which run longitudinally and which flatten out when the stomach is full. The mucus secreted by the goblet cells helps to lubricate the food

Box 20.6

Functions of the Stomach

The functions of the stomach are summarized in Box 20.7.

Functions of the stomach

- To churn up the food, breaking it up still further and mixing it with the secretions from the gastric glands. The part of the stomach just before the pyloric sphincter—the *pyloric antrum*—plays a large part in this movement, muscular contraction and relaxation sending some of the now liquid food through the sphincter into the small intestine and returning some to the body of the stomach for further mixing
- To continue the digestion of food by means of the gastric juice
- To secrete the intrinsic factor

Box 20.7

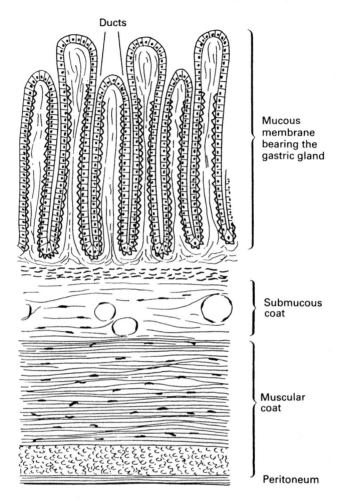

Ducts

Mucous membrane bearing the gastric gland

Submucous coat

Muscular coat

Peritoneum

Fig. 20.8 Section of gastric wall, highly magnified to show glands.

There are three types of cell in the mucosa of the stomach. *Mucous cells* secrete mucus which protects the mucous membrane from the action of the other gastric juices; *chief cells* secrete an enzyme known as pepsinogen, and in children another called rennin, and *oxyntic cells* secrete hydrochloric acid. The secretion of the gastric juice occurs reflexly in the same way as the saliva, causing a copious flow of the fluid before and during the taking of food. The gastric glands are also stimulated by an internal secretion or hormone, produced by the stomach, called *gastrin*, which passes into the circulation and when it reaches the gastric glands increases the production of gastric juice.

Gastric Juice
The constituents of gastric juice are listed in Box 20.8.

Contents of gastric juice

- Water, mineral salts and mucus
- Hydrochloric acid (HCl)
- Pepsinogen, which is converted by hydrochloric acid into the active enzyme pepsin. *Pepsin* turns proteins into peptides
- Rennin, which coagulates milk protein into casein, in which form it can be acted upon by pepsin. This is active in babies

Box 20.8

The juice makes the food more liquid and acid. Until it becomes acid, and this may take 15 to 30 minutes in the cardiac end of the stomach, which acts as a reservoir, the *amylase* of the saliva continues to act on the cooked starch. When the food is acid, *pepsin* and *rennin* act on the proteins and on caseinogen. The food is quickly acidified in the pyloric end of the stomach, where peristaltic action is very marked, so that it acts like a mill, churning the food up and mixing it with the gastric juice. The food remains in the stomach for $\frac{1}{2}$ hour to 3 hours or more, according to the nature of the food and the muscularity of the individual stomach. A meal rich in carbohydrate but containing little protein, such as tea, toast and cake, will leave the stomach in half an hour. A good mixed meal, like a typical dinner, will remain for $2\frac{1}{2}$ to 3 hours or more, though it may leave earlier or stay longer according to the tone and activity of the muscular coat.

The hydrochloric acid in the gastric juice serves several purposes (Box 20.9).

Purposes of gastric acid

- It gives the acid reaction required by the gastric enzymes
- It kills bacteria
- It controls the pylorus
- It stops the action of ptyalin
- It converts pepsinogen to pepsin

Box 20.9

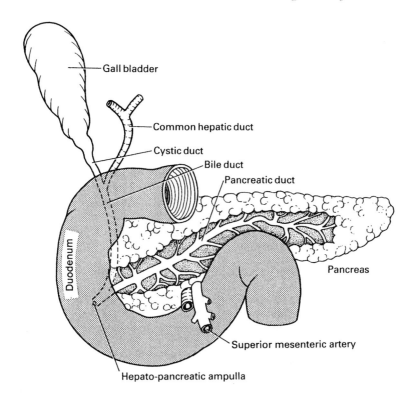

Fig. 20.9 The duodenum, pancreas and gall bladder.

The pylorus is normally contracted. When there is food in the stomach the gastric juice makes the contents gradually more and more acid at the pyloric end. When it reaches a certain degree of acidity the pylorus relaxes and a little food passes into the duodenum (Fig. 20.9). The acid food here causes the pylorus to close and the tone of the stomach wall drives food from the cardiac reservoir down to mix with the food in the pyloric end, making it less acid. Gradually the food in the duodenum is made alkaline and that in the pyloric end of the stomach again becomes more acid; this causes the pylorus to open again.

The churning action of the stomach serves to emulsify coarsely any fat which may be present and which the body heat will have melted. This converts the food into a greyish-white fluid called *chyme*.

THE SMALL INTESTINE

The small intestine is a convoluted tube extending from the pyloric sphincter to its junction with the large intestine at the ileo-caecal valve. It is about 6 metres in length and lies in the central and lower parts of the abdominal cavity, usually within the curves of the large intestine (see Fig. 20.1). The small intestine consists of the duodenum, the jejunum and the ileum.

The *duodenum* is a short curved portion of about 25 cm long and is the widest and most fixed part of the small intestine. It is roughly C-shaped and curves round the head of the pancreas. Ducts from the gall bladder and liver and the pancreas enter the medial aspect of the duodenum through the *hepato-pancreatic ampulla*, which is guarded by a sphincter-like muscle. The duodenum plays a central role in controlling digestion, which will be considered below. When the food from the stomach enters the duodenum, hormones are released which simultaneously stimulate the release of bile from the gall bladder and pancreatic juice from the pancreas. These hormones, *cholecystokinin*, *secretin* and another called *gastric inhibitory peptide*, also inhibit movement in the stomach and the secretion of gastric juice. In addition, the entry of food into the duodenum also stimulates the nervous *enterogastric reflex* which has an inhibitory effect on the stomach.

The *jejunum* is the name given to the upper two-fifths of the remainder of the small intestine and the lower three-fifths is called the *ileum*. Both are attached to the posterior abdominal wall by a fold of peritoneum called the mesentery (see Fig. 20.17).

The wall of the small intestine has the same four coats as the remainder of the alimentary tract.

1. Serous coat, formed of peritoneum
2. Muscular coat, with a thin external layer of longitudinal fibres and a thick internal layer of circular fibres
3. Submucous coat, containing blood vessels, lymph vessels and nerves
4. Mucous membrane lining

The mucous membrane lining has three special features (Box 20.10).

The small intestine contains considerable amounts of lymphoid tissue. *Solitary lymphatic follicles* (Fig. 20.12) are found throughout the mucous membrane but are most numerous in the lower part of the ileum. *Aggregated lymphatic follicles* form circular or oval patches large enough to be seen with the naked eye. These nodules deal

Features of the mucous membrane of the small intestine

- It is thrown into circular folds which, unlike the rugae of the stomach, are permanent and are not obliterated when the intestine is distended. They increase the area available for absorption (Fig. 20.10)
- It has a velvety appearance due to the presence of fine hair-like projections called *villi*, each containing a lymph vessel called a *lacteal*, and blood vessels (Fig. 20.11)
- It is supplied with *glands* of the simple, tubular type which secrete *intestinal juice*

Box 20.10

Fig. 20.10 Section of the small intestine, showing the puckered lining (circular folds).

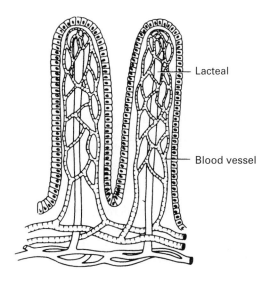

Lacteal

Blood vessel

Fig. 20.11 Villi containing blood vessels and lacteal.

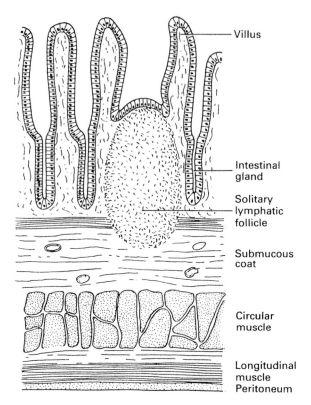

Fig. 20.12   Section of the wall of the small intestine (highly magnified).

with bacteria which may be absorbed from the intestine; they become inflamed as a complication of such diseases as typhoid fever.

Functions of the Small Intestine

The functions of the small intestine are the digestion and absorption of food.

Digestion

Digestion is carried out by the pancreatic juice, bile (which emulsifies fats) and intestinal juice (Table 20.2). There are three phases of digestion, which work together (Box 20.11). The *cephalic phase* occurs when we think of, see or smell food and it stimulates the release of gastric juice and begins movement in the stomach. This is the reason why our stomach apparently 'rumbles' when we are

Table 20.2 Digestive Processes

Source	Secretion	Changes
Saliva	Amylase	Cooked starch into maltose and dextrin
Gastric juice	Pepsin Rennin Hydrochloric acid	Proteins into peptones Coagulates milk (in babies) Stops action of amylase Converts pepsinogen to pepsin
Pancreatic juice	Trypsin Amylase Lipase	Proteins to polypeptides All starch to maltose and dextrin Fats to fatty acids and glycerol
Liver	Bile	Emulsifies fats
Intestinal juice	Enterokinase Peptidase Maltase Sucrase Lactase Lipase	Pancreatic trypsinogen to active trypsin Polypeptides to amino acids Maltose ⎫ Sucrose ⎬ to glucose Lactose ⎭ Fats to fatty acids and glycerol

hungry. The *gastric phase*, as the name suggests, occurs when food is in the stomach, and the presence of food here stimulates the release of gastric juice and movement. The *intestinal phase* occurs when food enters the duodenum and the secretion and movement in the stomach is inhibited by the mechanisms described above, which involve both hormonal and neural mechanisms.

The three phases of digestion

- The cephalic phase
- The gastric phase
- The intestinal phase

Box 20.11

The pancreatic and intestinal juices are secreted reflexly through sensations concerning food, and also through a hormone called *secretin* produced by the lining of the intestine, as gastrin is secreted by the stomach wall. The juices are alkaline and make the food alkaline in reaction. This finely emulsifies the fat.

The pancreatic juice consists of water, alkaline salts and three enzymes acting on three different foodstuffs (Box 20.12).

Contents of pancreatic juice

- **Trypsinogen**—turns peptones and proteins into amino acids when converted into active trypsin by enterokinase. If active trypsin were secreted by the pancreas, it would be able to digest the protein of the cells which form the gland and its ducts. Trypsin becomes active only when it mixes with the food and the intestinal juice within the bowel
- **Amylase**—turns starch, cooked and uncooked, into malt sugar (maltose)
- **Lipase**—splits fat into fatty acids and glycerol, after the bile has emulsified the fat, to increase the surface area

Box 20.12

The *bile* contains no enzymes, but is rich in alkaline salts, which serve to emulsify and saponify fats, i.e. make soaps from them.

The *intestinal juice* contains water, salts and enzymes (Box 20.13).

The enzymes in intestinal juice

- **Enterokinase**—converts trypsinogen secreted by the pancreas to active trypsin
- **Peptidase**—acts on peptones and turns them into amino acids
- **Maltase**—turns maltose into simple sugar, such as glucose
- **Sucrase**—turns cane sugar (sucrose) into simple sugar
- **Lactase**—turns lactose into simple sugar
- **Lipase**—completes the conversion of fats to fatty acids and glycerol

Box 20.13

These juices are mixed with the food by *peristalsis,* the muscular action of the wall of the small intestine. These movements may be seen on a film taken with the small intestine exposed by an abdominal incision. The contractions occur first at one place and then at another and are followed by relaxation, having a kneading or churning effect and bringing the mucous lining into close contact with the contents of the gut.

Absorption

The absorption of proteins, carbohydrates and fats takes place almost entirely through the villi in the small intestine. Very little food is absorbed from the stomach, as it is not yet sufficiently digested, or, if absorbable, e.g. glucose and water, does not stay in the stomach, but merely passes through it. *Proteins* in the form of amino acids and *carbohydrates* in the form of simple sugar are absorbed by the cells covering the villi and pass into the *blood capillaries*, being carried by the portal vein to the liver. *Fats* in the form of fatty acids and glycerol are absorbed by the cells covering the villi and built up again by them into droplets of fat. These pass into the lymph within the villi and are drained away by the *lymphatic capillaries* or lacteals, which are so named because the lymph they contain is made milk-like by the droplets of fat in suspension. The fat passes via the lymphatic vessels to the cisterna chyli and is carried up the thoracic duct into the blood stream.

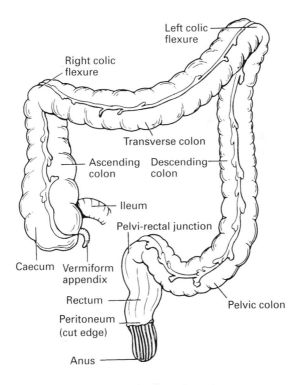

Fig 20.13   The large intestine.

THE LARGE INTESTINE

The large intestine extends from the end of the ileum to the anus (Fig. 20.13) and is about 1·5 metres long. It forms an arch which encloses most of the small intestine, and is divided into seven sections (Box 20.14).

Sections of the large intestine

- The caecum
- The ascending colon
- The transverse colon
- The descending colon
- The sigmoid colon
- The rectum
- The anal canal

Box 20.14

The *caecum* lies in the right iliac fossa. It is a dilated area which has a blind lower end but is continuous above with the ascending colon and where one passes into the other the ileum opens into the caecum through the *ileo-caecal valve* (Fig. 20.14). This valve is a sphincter and prevents the caecal contents passing back into the ileum. Taking food into the stomach initiates contraction of the duodenum and the rest of the small intestine, followed by the passage of the contents of the ileum into the caecum through the ileo-caecal valve. This is called the *gastro-ileal reflex*.

The *vermiform appendix* is a narrow blind-ended tube which opens out of the caecum about 2 cm below the ileo-caecal valve. It is usually about 9 cm long, though it can vary from 2 to 20 cm in length and can occupy a variety of positions within the abdomen. The submucous coat of the appendix contains considerable amounts of lymphoid tissue.

The *ascending colon* (Fig. 20.15) is about 15 cm in length and is narrower than the caecum. It ascends the right side of the abdomen to the under surface of the liver where it bends forwards and to the left at the *right colic flexure*.

The *transverse colon* is about 50 cm in length and passes across the abdomen to the under surface of the spleen in an inverted arch. Here it curves sharply downwards at the *left colic flexure*.

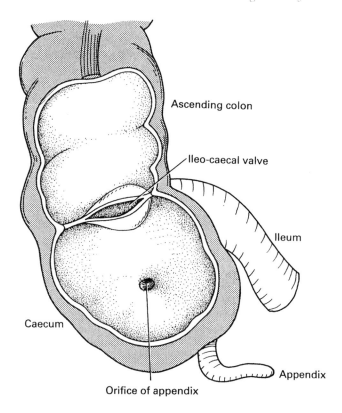

Ascending colon

Ileo-caecal valve

Ileum

Caecum

Appendix

Orifice of appendix

Fig. 20.14 The caecum with the appendix.

The *descending colon* is about 25 cm in length and passes down the left side of the abdomen to the inlet of the lesser pelvis, where it becomes the *sigmoid colon*.

The *sigmoid colon* forms a loop which is about 40 cm in length and lies within the lesser pelvis.

The *rectum* is continuous above with the sigmoid colon. It is about 12 cm long and passes through the pelvic diaphragm to become the anal canal.

The *anal canal* passes downwards and backwards to end at the anus. At the junction of the anus and the rectum the unstriped circular muscle becomes thickened to form the *internal anal sphincter* which surrounds the upper three-quarters of the anal canal. The *external anal sphincter* surrounds the whole length of the anal canal and it is the tone of these sphincters which keep the anal canal and the anus closed. The external sphincter can be contracted voluntarily to close the anus more firmly.

Fig. 20.15   The colon. (1) Caecum with appendix; (2) ascending colon; (3) right colic flexure; (4) transverse colon; (5) left colic flexure; (6) descending colon; (7) sigmoid flexure; (8) beginning of the rectum.

The wall of the large intestine has the same four layers as the remainder of the alimentary tract (Box 20.15).

The layers of the large intestine

- The outer **serous layer** of peritoneum
- The **muscular layer** of external longitudinal and internal circular fibres. The longitudinal fibres form a continuous layer, but in some places the layer is thickened to form three bands called *taeniae coli*. The bands are shorter than the

other coats of the large intestine and produce the typical puckered or sacculated appearance. The sacculations are called the *haustrations*

- The **submucous layer**
- The lining of mucous membrane

Box 20.15

Functions of the Large Intestine

The functions of the large intestine

- To absorb water and salts
- The excrete faeces

Box 20.16

The material which enters the large intestine (Fig. 20.16) consists of water, salts, very little food material, as this has been digested and absorbed in the small intestine, cellulose, which is indigestible, and bacteria. The bacteria are very numerous; although they are largely killed in the stomach, the alkaline reaction, food, warmth and moisture in the small intestine encourage their growth here. The material is in a very fluid state. In the colon water and salts are quickly absorbed (Box 20.16) so that the fluid is rapidly turned into a paste containing the cellulose and bacteria, many of which die from lack of water and food. This paste forms the faeces, which consists of a fluid and solid part, the latter being roughly 50 per

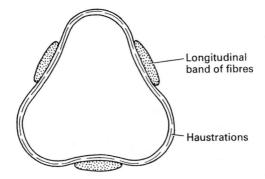

Fig. 20.16 Cross-section of the large intestine.

cent cellulose and 50 per cent dead bacteria. At intervals mass movements propel the faeces into the rectum, from which it is excreted.

Movements of the colon are similar to those seen in the small intestine but peristalsis occurs less frequently. A very strong wave of peristalsis occurs three to four times a day and moves material on into the pelvic colon.

Defaecation, or the passage of faeces, occurs because of the movement of food residue from the pelvic colon into the rectum, which becomes distended. This distenstion causes a reflex contraction of the rectal muscles which tends to expel the contents through the anus, though this depends on the relaxation of the external anal sphincter. Defaecation is therefore a reflex action which can be

Fig. 20.17 The peritoneum.

inhibited voluntarily. The gastro-ileal reflex also causes emptying of the rectum. If defaecation is delayed the sensation of fullness in the rectum passes off, more water is absorbed from the faeces through the rectal wall and constipation may occur.

THE PERITONEUM

The peritoneum (Fig. 20.17) is a serous membrane and in the male is a closed sac lining the abdomen. In the female the free ends of the uterine tubes open into the peritoneal cavity. The part which lines the abdominal wall is named the *parietal portion* of the peritoneum; that part which is reflected over the organs is called the *visceral portion*. The two layers are in contact with each other but the potential space between them is named the *peritoneal cavity*. This cavity consists of the greater peritoneal sac and the lesser peritoneal sac.

Other specially named areas are the *greater omentum*, a double fold of peritoneum which descends from the lower border of the stomach and loops up again to the transverse colon. This helps to prevent the spread of infection from the organs into the peritoneum. The *lesser omentum* is a fold which extends to the liver from the lesser curvature of the stomach and the duodenum. The *mesentery* is a broad, fan-shaped fold of peritoneum connecting the coils of the small intestine to the posterior abdominal wall.

The functions of the peritoneum are summarized in Box 20.17.

Functions of the peritoneum

- To prevent friction as the abdominal organs move on one another and against the abdominal wall, as its free surfaces are always moist with serum and are smooth and glistening
- To attach the abdominal organs to the abdominal wall, except in the case of the kidneys, duodenum and pancreas, which lie behind it. The ascending and descending colon are covered only on their anterior surface by peritoneum. This means that only the transverse or sigmoid colon can be brought on to the anterior abdominal wall for a colostomy operation
- To carry blood vessels, lymphatics and nerves to the organs; these run between the two folds of the membrane
- To fight infection, as the peritoneum contains many lymphatic nodes

Box 20.17

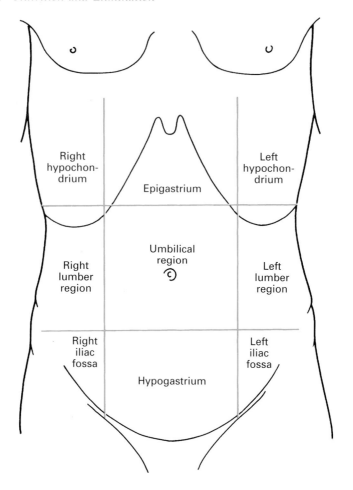

Fig. 20.18   Regions of the abdomen.

REGIONS OF THE ABDOMEN

For the purposes of description, the abdomen is divided into nine regions by two transverse and two upright lines (Fig. 20.18). In describing the position of organs, the regions in which they lie may be used, e.g. the stomach lies in the left hypochondriac, epigastric and umbilical regions. The kidneys are in the right and left lumbar regions respectively. The caecum is in the right iliac fossa. The bladder rises when full into the hypogastric region.

Chapter 21
The Liver, Biliary System and Pancreas

Due to their anatomical proximity and their related functions in digestion, the liver, the biliary system and the pancreas are usually considered together. This chapter looks at the structure of the liver and its function, particularly in the production of bile. The way in which the biliary system concentrates and subsequently releases bile is described and, finally, the structure and secretions of the pancreas are considered. The function of the pancreatic enzymes is the subject of Chapter 23.

THE LIVER

The liver is the largest gland in the body. It is situated in the upper right part of the abdominal cavity, occupying almost all of the right hypochondrium and fitting under the diaphragm. It has two main *lobes*, the right lobe being much larger than the left (Fig. 21.1). The right lobe lies over the right colic flexure and the right kidney and the left lobe over the stomach.

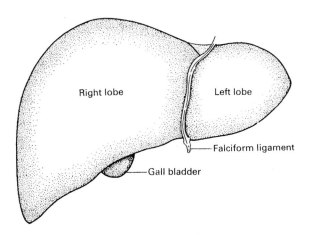

Fig. 21.1 The liver, general appearance.

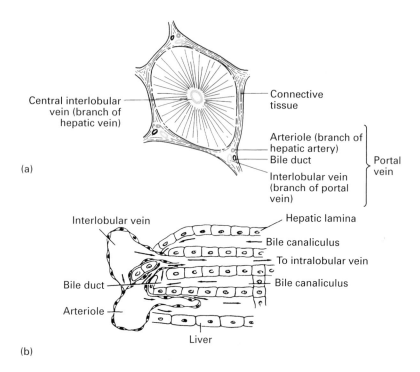

Fig. 21.2   (a) A liver lobule; (b) the scheme of blood vessels and bile ducts.

Structure of the Liver

The liver consists of a large number of *hepatic lobules* which appear hexagonal in shape (Fig. 21.2). Each is about 1 mm in diameter and has a small central intralobular vein (a tributary of the hepatic veins). Around the edges of the lobules are the *portal canals* each containing a branch of the *portal vein* (interlobular vein), a branch of the *hepatic artery* and a small *bile duct*. These three structures together are known as the *portal triad*.

The lobules are composed of *liver cells* which are large cells with one or two nuclei and fine granular cytoplasm. The liver cells are arranged in sheets, one cell in thickness, called *hepatic laminae*. These laminae are arranged irregularly to form walls with bridges of liver cells connecting adjacent laminae. Between the laminae are spaces which contain small veins with many anastomoses between them and small bile ducts called *canaliculi*.

In the liver the portal vein brings blood rich in foodstuffs from the alimentary tract and the hepatic artery brings blood rich in oxygen from the arterial system. These divide up into smaller vessels and provide a capillary network among the liver cells forming the hepatic laminae. This capillary network then drains into the small veins in the centre of each lobule which supply the hepatic vein. These vessels carry away the blood from both the portal capillaries and the deoxygenated blood which has been brought to the liver by the hepatic arteries as oxygenated blood.

Functions of the Liver

The functions of the liver can be divided into three sections: metabolic, storage and secretory (Box 21.1).

Functions of the liver

Metabolic functions
- Stored fat is broken down to provide energy; this process is called *desaturation*
- Excess amino acids are broken down and converted to urea
- Drugs and poisons are de-toxicated
- Vitamin A is synthesized from carotene
- The liver is the main heat-producing organ of the body
- The plasma proteins are synthesized
- Worn-out tissue cells are broken down to form uric acid and urea
- Excess carbohydrate is converted to fat for storage in the fat depots
- Prothrombin and fibrinogen are synthesized from amino acids
- Antibodies and antitoxins are manufactured
- Heparin is manufactured

Storage functions
- Vitamins A and D
- Anti-anaemic factor
- Iron from the diet and from worn-out blood cells
- Glucose is stored as glycogen and converted back to glucose in the presence of glucagon as required

Secretory functions
- Bile is formed from constituents brought by the blood

Box 21.1

The Formation of Urea

The amino acids derived by the process of digestion from the protein food we eat are absorbed by the villi of the small intestine and brought by the portal vein to the liver. The amino acids required to make good the wear and tear of tissue and produce its growth are allowed to pass straight through the liver into the blood stream. Others are used to form the blood proteins. Any excess protein or second-class protein which is unsuitable for tissue building is broken up in the liver to form: (a) body fuel, composed of carbon, hydrogen and oxygen; and (b) urea, a compound containing the nitrogen present in all proteins, which is incombustible, and therefore useless unless needed for the building of tissues. This urea is a soluble substance which the blood stream carries from the liver to the kidneys for excretion from the body.

The Secretion of Bile

The bile is a thick greenish-yellow fluid secreted by the liver cells. It is alkaline in reaction. The liver secretes on average about a litre of bile per day. The bile consists of water, bile salts and bile pigments. The bile salts give the bile its alkaline reaction and include both organic and inorganic salts. Among the former is the substance cholesterol, which is the main ingredient of the typical gallstone. The bile pigments are derived from the haemoglobin of the worn-out red corpuscles which are excreted from the body by this channel, giving the normal colour to the faeces: they are also present in the blood and colour the urine.

The uses of bile are listed in Box 21.2.

Functions of bile

- It **helps to emulsify and saponify fats** in the small intestine by its alkalis; in this way the surface area is increased and the action of enzymes is increased
- It **stimulates peristalsis in the intestine**, so that it acts as a natural aperient
- It is **a channel for excretion** of pigments and toxic substances from the blood stream, such as alcohol and other drugs
- It acts as **a deodorant to the faeces**, lessening their offensive odour. It is suggested that this may be due merely to the fact that a lack of bile means poor digestion of fat, so that fat remains in the intestine in excess, coating the other foods and

preventing their digestion and absorption. As a result undigested protein is attacked by bacteria and, decomposing, produces an excess of sulphuretted hydrogen, the gas which causes the smell of abnormal faeces, foul drains and rotten eggs

Box 21.2

THE BILIARY SYSTEM

The elements of the biliary system are listed in Box 21.3.

Components of the biliary system

- The **right and left hepatic ducts** from the liver which unite to form the **common hepatic duct**
- The **gall bladder**, which acts as a reservoir for bile
- The **cystic duct**, leading from the gall bladder
- The **bile duct** formed by the junction of the common hepatic and cystic ducts

Box 21.3

The *gall bladder* is a pear-shaped organ situated on the under surface of the right lobe of the liver. From it the *cystic duct*, which is about 3 to 4 cm long, passes backwards and downwards to join the *common hepatic duct* and together they form the *bile duct* (Figs. 21.3 and 21.4). If the bile secreted by the liver is not required immediately for purposes of digestion it passes up the cystic duct into the gall bladder where it is both stored and concentrated. The capacity of the gall bladder is between 30 and 60 ml but because of its capacity to absorb water the bile it contains becomes increasingly concentrated. When fatty food enters the duodenum the sphincter at the entrance to the bile duct relaxes and bile stored in the gall bladder is driven into the intestine by contraction of the walls of the gall bladder (Fig. 21.5).

From the list of its functions it will be seen that the liver is essential to life. It is, however, able to undertake more work than is usually demanded of it and a considerable part may be destroyed by disease before death occurs from liver failure.

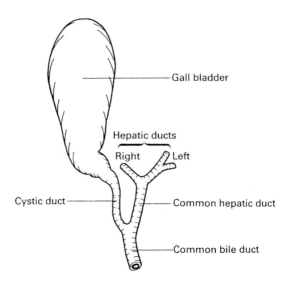

Fig. 21.3 The gall bladder and its ducts.

Fig. 21.4 X-ray showing the normal gall bladder filled with an opaque drug (e.g. Opacol).

Fig. 21.5 The gall bladder emptying after a fatty meal; the opaque medium shows the bile ducts.

THE PANCREAS

The pancreas is a soft greyish-pink gland, 12 to 15 cm in length, lying transversely across the posterior abdominal wall (see Fig. 20.9) behind the stomach. The *head* of the gland lies within the curve of the duodenum and the *tail* extends as far as the spleen. The *body* lies between these two. The *pancreatic duct* lies within the organ. It begins with the junction of the small ducts from the pancreatic lobules in the tail of the pancreas and runs from left to right through the gland, receiving ducts all the way. At the head of the pancreas the pancreatic duct is joined by the bile duct and they usually open together into the duodenum at the *hepato-pancreatic ampulla*, though occasionally there are two separate ducts.

The pancreas is composed of lobules, each of which consists of one of the tiny vessels leading to the main duct and ending in a number of alveoli. The alveoli are lined with cells which secrete enzymes called *trypsinogen, amylase* and *lipase* (Box 21.4)

Pancreatic enzymes

- **Trypsinogen** is converted into active **trypsin** by enterokinase, an enzyme secreted by the small intestine. In its active form trypsin converts peptones and proteins into amino acids
- **Amylase** turns starches, both cooked and uncooked, into maltose (malt sugar)
- **Lipase** splits fats into fatty acids and glycerol after the bile has emulsified the fat which increases the surface area

Box 21.4

Between the alveoli, collections of cells are found forming a network in which there are many capilliaries (Fig. 21.6). These cells are called *interalveolar cell islets* (islets of Langerhans) and they secrete a hormone which passes directly into the blood stream. The pancreas therefore has both a digestive and an endocrine function. Each islet consists of two types of cell designated alpha and beta. Alpha cells form about 25 per cent of the total number of islets and they produce a hormone called *glucagon* which is secreted in response to a fall in blood sugar. Glucagon stimulates the conversion of glycogen to glucose, thus raising the blood sugar level. Beta cells form the remaining 75 per cent of the islets and they secrete the hormone *insulin* in response to a rise in

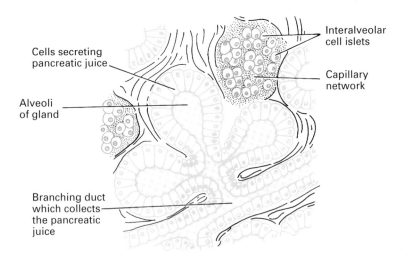

Fig. 21.6 The minute structure of the pancreas, showing the saccules which secrete the pancreatic juice and the islet cells which secrete insulin.

blood sugar level, e.g. after a meal. Insulin lowers the blood sugar level by stimulating the conversion of glucose to glycogen for storage and by increasing cellular uptake of glucose. It will be apparent therefore that the blood sugar level is maintained by a balance between the two hormones because they both affect carbohydrate metabolism. As the metabolism of proteins and fats are closely related to carbohydrate metabolism a disturbance in one will affect the others.

A deficiency of insulin results in diabetes mellitus. The blood sugar level rises above the renal threshold and glucose is lost in the urine. Because the cells cannot utilize glucose there is an accumulation of ketone bodies as a result of the breakdown of fatty acids which causes acidosis and may lead to coma and death if untreated.

Chapter 22
Nutrition

Good health is dependent on satisfactory nutrition, and this is in turn dependent on plentiful supplies of the foods which are necessary to healthy life. Food is one of the essential needs of the body. Substances which can serve as food for the body are those which it can use as *fuel* for combustion, or *building material* for the repair and growth of tissues. Fuel is required to produce the energy for the activities of every living thing and to maintain the heat at which the individual exists. Building materials are necessary to repair the body tissues, since they are constantly active and being worn out by their activities; in addition, in infants and children, extra building material is required to build up the new tissues needed for the processes of growth. Fuel supplies and building materials alone are, however, not enough. Certain other substances are necessary to enable the tissues to use the building materials and the fuel supplies, and these are known as vitamins.

There are six essential foodstuffs with which the body must be constantly supplied through the foods we eat (Box 22.1). Every article of food contains one or more of these foodstuffs and is only of value as food because it does contain them. The proteins, water and salts are the body-building foods; the carbohydrates and fats are essentially the fuel foods, though the body can and does use protein also as fuel, if more is taken in than is required for body building, or if there is a lack of other fuel, as in starvation. The vitamins and certain salts act as regulators of tissue activity, so that, although vitamins are of no use either as fuel or as body builders, the nutrition of the tissues suffers if they are not present in sufficient quantities in the food supply, and diseases appear which can both be prevented and cured by ensuring that the vitamins are present in satisfactory quantities in the diet.

To be of use to the body these foodstuffs must be digested and absorbed. Food must therefore be of such a nature that it can be digested, i.e. broken down by digestive juices into substances that can pass into the blood stream, and be carried to the various tissues for their use. Proteins, carbohydrates and fats are complicated compounds found in plant and animal matter, and require digestion.

Water and mineral salts are simple inorganic substances; they

---

Essential foodstuffs

- Proteins
- Carbohydrates
- Fats
- Water
- Mineral salts
- Vitamins

Box 22.1

---

can be absorbed, therefore, without digestion and they enter into the composition of all animal and plant matter. In fact, all living matter consists largely of water. The inorganic salts, absorbed from the soil or water, are built up by living things into organic salts which are an essential part of all plants and animals (see chapter 2 for more details of the chemical structure of nutrients).

PROTEINS

Proteins are the most complicated of the foodstuffs. They consist of carbon, hydrogen, oxygen, nitrogen and sulphur and usually phosphorus. They are often spoken of as the nitrogenous food-stuffs, as they are the only ones which contain the element nitrogen. They are essential for the building up of living protoplasm, as this also consists of the elements carbon, hydrogen, oxygen, nitrogen and sulphur. They are found in both animal and plant matter, but the animal proteins are the most valuable to the human body as building material, since they are similar to human protein in composition. On the other hand, plant proteins are cheaper; they are more useful as body fuel than as body builders, but do provide at a lower cost some of the amino acids that the body needs for tissue building. The sources of protein are listed in Box 22.2.

All proteins are made up of simpler substances known as *amino acids*. There are about 20 of these amino acids, but each protein contains some of these only. The amino acids are like letters from which many words can be made, each word being a different combination of letters. The protein of each different type of animal or plant is a different combination of amino acids. Ten essential amino acids are found in human protein. These are amino acids which the body cannot build up for itself. Proteins which contain all ten are called *complete proteins*, e.g. albumin, myosin, casein.

Sources of protein

Animal
- Eggs, containing albumin
- Lean meat, containing myosin
- Milk, containing caseinogen and lactalbumin
- Cheese, containing casein

Plant
- Wheat and rye, containing gluten
- Pulse foods (e.g. peas and beans) containing legumin

Box 22.2

Proteins which do not contain all the ten are called *incomplete proteins*, e.g. gelatin, which is contained in all fibrous tissue, and is extracted from bone and calves' feet in the making of soups and jelly. Animal proteins, such as those of eggs, milk and meat, not only contain all the ten amino acids the body needs, but contain all of them in good proportion; these are called *first-class proteins* and are the best building material for the body tissues. Plant proteins, such as gluten and legumin, contain only slight quantities of one or more of the ten amino acids essential to the body, and are therefore called *second-class proteins*, as they are not such good building material. Some first-class animal protein should always be included in the diet.

CARBOHYDRATES

Carbohydrates include *sugar* and *starch*. They consist of carbon, hydrogen and oxygen, and always contain hydrogen and oxygen in the same proportions as in water, i.e. twice as much hydrogen as oxygen. They are the chief sources of body fuel, being most easy to digest and absorb and most readily burnt in the tissues, being broken down into carbon dioxide and water. They are obtained chiefly from plant foods (Box 22.3).

Sources of starch and sugar

Sources of starch
- Cereals, e.g. wheat, rice, barley
- Tubers and roots, e.g. potatoes and parsnips

- Pulse foods, e.g. peas, beans and lentils

Sources of sugar
- Sugar cane (sucrose)
- Beetroot and all sweet vegetables and fruits, e.g. grape sugar or glucose
- Honey
- Milk, containing lactose or milk sugar

Box 22.3

Sugars are of three types (Box 22.4). Starch differs from sugar in that it is insoluble in water. Plants store sugar in the form of starch to prevent it from escaping in solution into the water in the soil in which they live. (Formula for starch $n(C_6H_{10}O_5)$, a polysaccharide; n stands for different numbers in the different starches of various plants.) All carbohydrates are reduced to monosaccharides before they can be absorbed from the digestive tract.

Types of sugar

- **Simple sugars** (monosaccharides), such as glucose or grape sugar (formula $C_6H_{12}O_6$)
- **Complex sugars** (disaccharides), such as cane sugar or sucrose, and milk sugar or lactose (formula $C_{12}H_{22}O_{11}$)
- **Polysaccharides**, the most complex carbohydrates; starches such as potatoes, cereals and root vegetables

Box 22.4

FATS

Fats, like carbohydrates, consist of carbon, hydrogen and oxygen, but do not contain as much oxygen in proportion to the hydrogen present. They also serve as body fuel. They are the best source of fuel from the point of view that 1 g of fat produces twice as much energy as 1 g of sugar. On the other hand, they are not so easy to digest and absorb and not so readily burnt in the tissues. Fats can only be completely broken down to form the final products of metabolism, carbon dioxide and water, if they are metabolised with sugar. If sufficient sugar is not metabolised with them, the breakdown of fat is incomplete and acid or acetone bodies are formed in the tissues. These acetone bodies cause fatigue in the muscles and if present in large quantities alter the pH of the blood, causing a condition known as acidosis which may lead to coma and death. Severe acidosis is only likely to occur in diabetic patients, who are unable to metabolise sugar, and in starvation, when the small quantity of sugar which the body can store has been used up and the large quantities of fat which the body can store are forming the main source of body fuel. Fats are obtained from both animal and plant matter; the main sources are listed in Box 22.5.

Sources of fat

Animal
- Fat meat and fish oils
- Butter
- Milk and cream

Plant
- Nut oil, contained in margarine
- Olive oil

Box 22.5

Fats are compounds of glycerol and fatty acids. Different fats contain different acids, e.g. fat meat contains stearic acid, butter contains butyric acid, olive oil contains oleic acid, and so on.

Animal fats are more expensive than plant fats but are more valuable as food becuase they contain the fat-soluble vitamins A and D provided the animals have been exposed to sunlight and not kept in stalls. However, plant fat can be made equally valuable by the addition of Vitamin D, and all margarine, whether manufactured from fish oil or plant oil, is so treated today to ensure the supply of these vitamins. In this way the population has been protected from deficiency diseases due to lack of these fat-soluble vitamins.

WATER

Water is a simple compound of hydrogen, 2 parts, and oxygen, 1 part. It forms two-thirds of the body and is present in most of the foods we eat. Lean meat is three-quarters water, milk contains 87 per cent water, while cabbage contains as much as 92 per cent water. In addition to the water contained in food the body needs 2 to 3 litres of water every day. Water is required for many purposes (Box 22.6).

The purposes of water in the body

- The building of body tissues and body fluids
- The excretion of waste products
- The making of digestive and lubricating fluids
- The cooling of the body by the evaporation of sweat

Box 22.6

Fluid and electrolyte balance is discussed in Chapter 24.

ELECTROLYTES AND MINERALS

Salts and the ions produced from them are constantly taken into and lost from the blood. Salts are contained in all foodstuffs, but added salt, used in cooking and in eating, is necessary to maintain the normal electrolyte balance and replace the salt lost in urine and sweat; 3 to 4 g of sodium chloride, or common salt, is needed daily.

The quantities of the various salts are estimated as millimoles/ litre (mmol/l) of the ions they produce. In plasma there are normally 155 mmol/l of negatively charged ions, balanced by 155 mmol/l of positively charged ions. The bulk of these are provided by chloride (102 mmol/l) and sodium (143 mmol/l). These figures are given for reference, as the nurse will see them in laboratory reports to check for deficiencies or excess of sodium, potassium chloride, bicarbonate, etc., in the blood. These may result from deficiency in intake or excessive or reduced loss, especially in abnormal conditions affecting the kidney, or from heavy sweating or fever.

MINERAL SALTS

Salts are produced by the action of an acid on a mineral, which is termed the base of the salt, e.g. sodium chloride or common salt, produced by the action of hydrochloric acid on sodium; calcium lactate, produced by the action of lactic acid on calcium. Various salts and the ions split off from them are required by the body, every tissue and fluid in the body containing them. Chlorides, carbonates and phosphates of sodium, potassium and calcium are particularly important. Salts are required for body building and are also regulators of tissue activity. They provide the electrolytes, capable of carrying positive and negative electric charges in the body fluids (see Chapter 1). A correct balance of electrolytes is essential for normal activity in the body tissues and fluids.

Sodium is present in all tissues; the body contains it in the form of sodium chloride to the extent of 9 g to 1 litre (0.9 per cent). It is present in the same concentration in all tissue fluid. Sodium carbonate and sodium phosphate are also always present in the blood and tissues. The carbonates give the alkaline reaction to the blood and form the alkali reserve, neutralizing the carbonic acid produced by fuel combustion. The phosphate is the carrier of the acids produced by the breakdown of the body building foods and carries them to the kidneys, by which they are excreted. It is obtained from our food, particularly animal foodstuffs, and from the rock salt used in cooking.

Potassium is present in all tissue cells, where it replaces the sodium of blood and tissue fluid as the base and the source of the positively charged ions. It is obtained from our food, particularly plant foodstuffs.

Calcium is present in all tissues, particularly in bone, in teeth, and in blood, and is necessary for normal functioning of nerves. It is obtained chiefly from milk, cheese, eggs and green vegetables, and to some extent from hard water, though this inorganic form is less valuable than the organic salts in plant and animal foods. Adults require 400 to 500 mg daily.

Iron is essential for the formation of the haemoglobin of red blood cells. It is obtained from green vegetables, particularly spinach and cabbage, egg yolk and red meats. Men require 10 mg daily; women 10 to 15 mg.

Phosphorus is also needed for the building of the body tissues. It is obtained from the yolk of egg, milk and green vegetables.

Iodine is required for the formation of the secretion of the thyroid gland. It is obtained from seafood and is present in green vegetables, which have derived it from the fine spray of sea water in the air which is carried inland by winds.

Calcium, iron and iodine are the only minerals likely to be insufficient. The others are present in adequate amounts in the diet.

VITAMINS

Vitamins are substances which are also essential to normal health, although they are of no value to the body as fuel nor as building material. Without them diseases occur which are known as deficiency diseases. They are present in small quantities in living foodstuffs and are required in minute traces only each day. Their discovery dates back to the period following the 1914–18 war, and as their composition was at first unknown they were named after the letters of the alphabet. They can today be made synthetically to a large extent.

There are numbers of vitamins now identified, but experimental work still continues. The chief vitamins are: vitamin A, the vitamin B complex, vitamin C, vitamin D, vitamin E and vitamin K.

Vitamin A

Vitamin A is present in animal fats, being fat-soluble. In carrots and green vegetables, and in all yellow fruits a substance called *carotene* is found, which, like the green colouring matter chlorophyll, is formed under the influence of sunlight. This is a precursor of vitamin A, and animals, including man, can turn the carotene present in their food into vitamin A within their bodies. Lack of vitamin A causes stunted growth and lowered resistance to infection. The mucous membranes particularly are unhealthy and an easy prey to bacteria when the diet is deficient in it. The conjunctiva of the eye is affected and a form of conjunctivitis occurs, known as *xerophthalmia*, in which the conjunctiva loses its transparency, and a horniness or cornification develops in it. The retina is also affected and night blindness (i.e. inability to see at night) develops—a symptom which became apparent in the 'black-out' during the war, when some people found they could not see as well as others in the dark, because their diet had not been satisfactory.

Vitamin B

The vitamin B complex consists of a number of factors, although originally thought to be a single substance. These factors were confused because they had a similar distribution, being found particularly in the husks and germs of cereals and pulses, and in yeast and yeast extracts. They are also present to a lesser extent in vegetables, fruit, milk, eggs and meat. White flour and the bread, cakes and pastries made from it do not contain these factors, nor do polished rice and barley. Hence brown bread and wholemeal flour have a food value that white bread and flour lack and where these latter form a large part of the diet a subnormal state of health may be present, or even deficiency disease, as occurred in many of the prison camps during the last war, especially in the Far East. The chief factors in the vitamin B complex are:

Vitamin B (*aneurine* or *thiamine*). This is essential for carbohydrate metabolism and controls the nutrition of nerve cells; a marked deficiency of it leads to beri-beri, in which there is inflammation of the nerves, causing paralysis and a loss of tone and activity in the intestinal muscles, with constipation, while the patient complains of loss of appetite and a burning sensation in the feet.

Vitamin B$_2$ (*riboflavine*). This is essential for proper functioning of cell enzymes.

Vitamin B$_3$ (*nicotinic acid*). This is essential for carbohydrate metabolism. A lack of this causes pellagra, which leads to skin eruptions, gastro-intestinal changes and mental changes.

Vitamin B$_6$ (*pyridoxine*). This is believed to be necessary for protein metabolism.

Vitamin B$_{12}$ (*cyanocobalamin*). This is the anti-anaemic substance or factor absorbed by the villi of the small intestine and stored in the liver. It is satisfactorily absorbed only in the presence of an intrinsic factor produced by the lining of the stomach and hydrochloric acid. Vitamin B$_{12}$ is essential for the proper development of red cells in the red bone marrow. Lack of it, or of its absorption, causes pernicious anaemia.

Folic acid. This is part of the vitamin B complex. It is also necessary for the maturation of red blood cells.

Vitamin C

Vitamin C (ascorbic acid) is water-soluble and is found in fresh fruit, particularly citrus fruits (oranges, grapefruit and lemons) and in green vegetables and potatoes. It is important in tissue respiratory activity, wound repair and resistance to infection, and affects the condition of capillary walls, which become abnormally fragile if it is not present in plentiful supply in the diet. Lack of it causes scurvy, so that it is called the antiscorbutic vitamin. It is particularly readily destroyed by heat, hence some fresh fruit or salad should be included in the daily diet. Cabbage and other greens provide a very rich source of vitamin C and are comparatively cheap. Eaten raw, as salads, in a finely shredded form, they are very palatable and most valuable, provided the cabbage is fresh and crisp. Cooking for a long period of time lessens the content of vitamin C, and for this reason cabbage should be finely shredded and cooked in boiling water only until tender, i.e. for 10 to 15 minutes.

Under normal circumstances the eating of a good mixed diet, containing plenty of fresh fruit and vegetables, makes the taking of vitamin C in tablet form unnecessary and inadvisable.

Vitamin D

Vitamin D is fat-soluble and is found with vitamin A in animal fats, provided the animal has been in the sun. Cod-liver and halibut-liver oil are very rich in it. Halibut oil is more expensive as the supplies are less plentiful, but it is richer in content, so that it is useful for those who cannot digest cod-liver oil. Vitamin D can also be made in the skin by the action of ultraviolet rays on the ergosterols present. Further, it can be manufactured by subjecting the sterols in fats to the action of ultraviolet rays. The product of this process is named *calciferol* and this can be obtained in tablet form. It is identical with the vitamin D present in animal fat and can be used in illness and in emergencies as a substitute for it. It is essential for the development of bone and teeth, affecting the use of calcium and phosphorus in the body. Lack of vitamin D causes rickets, so that it is known as the antirachitic vitamin. This disease was common among the poor of industrial areas who ate margarine in place of butter, especially in countries like the British Isles, where there was not a great deal of sunshine. However, modern methods of manufacturing the vitamin have made it possible to add it to margarine, with the result that rickets as a disease has practically disappeared in this country, since it has become compulsory for all makers of margarine to ensure that it contains the vitamin.

Vitamin E

Vitamin E is present in vegetable oils. It is present in cereals and has been shown to be essential for reproduction in rats. Little is known of its importance in human beings.

Vitamin K

Vitamin K is fat-soluble and can be obtained from green vegetables and liver. Persons who have had deficiency in their diet show a tendency to haemorrhage, since vitamin K helps to form the prothrombin in the blood. It is synthesized in the intestines by bacterial action.

The quantities of the various vitamins required in the normal diet are estimated in milligrams; pure vitamins can be given in measured quantities on the doctor's prescription.

ROUGHAGE

The food should also contain cellulose. This is indigestible and therefore remains in the bowel and stimulates it to empty itself. It is called roughage. Cellulose forms the fibrous part of plant foods and is present in all green vegetables, in fruits, the skin of beans and peas, and in wholemeal flour and bread. These foods are therefore essential to prevent constipation.

DIET

Precise nutritional guidelines for dietary intake which are suitable for each individual are impossible to give. Generally speaking, a daily intake of 2500 kcal is required, but this requirement may rise considerably for people who are very active. It is the way in which these calories are obtained which is presently the focus of dietary advice, and in 1989 the National Academy of Sciences (NAS) of the USA published a reasonable set of guidelines with respect to this aspect of diet and with respect to protein intake. If adequate attention is paid to these aspects of diet then an adequate intake of vitamins and minerals is usually assured.

The NAS recommended that fat intake should be less than 30% of calorie intake and that the intake of saturated fats (which may be associated with cardiovascular disease) should be less than 10% of calorie intake. This can be achieved by substituting fish, poultry without skin, lean meats and low-fat dairy products for fatty meats and whole-milk dairy products. In addition, vegetables, fruit and cereals can be consumed and the intake of oils, fats, eggs and fried foods can be limited.

Specifically, it is recommended that five servings (one serving being roughly equivalent to half a cup) of a combination of vegetables and fruit can be consumed. Carbohydrates should be consumed in complex forms such as starch rather than in refined forms such as sucrose and glucose. Protein intake should be about

0.8 gram per kilogram of body weight and no more than 1.6 gram per kilogram of body weight.

Salt (sodium chloride) intake should be no greater than 6 gram daily and this can be achieved by not adding salt to food at the table and limiting its use in cooking. A potassium intake, from an adequate intake of fruit and vegetables, is 3.5 gram daily, and an adequate intake of calcium can be achieved by the use of non-fat dairy products.

The use of supplements to the diet is not recommended, and it is emphasized that the best way to maintain a desirable body weight is to balance dietary intake with adequate exercise.

Chapter 23
Digestion and Metabolism

The anatomy of the digestive tract was considered in Chapter 20. The aim of the present chapter is to consider the chemical breakdown (digestion) of individual foodstuffs and their subsequent use in metabolism to provide energy and structural components for the body.

The digestion of the major sources of nutrition carbohydrate, fat and protein is dealt with in the following sections, and the inter-related nature of their metabolism in a subsequent section (Figs 23.1 and 23.2).

DIGESTION OF CARBOHYDRATES

Carbohydrate is present in the diet in several forms—as polysaccharides, disaccharides and monosaccharides. The polysaccharides

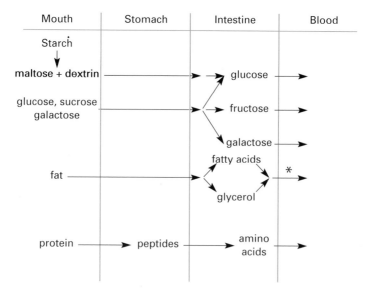

* Via lymphatic system

Fig. 23.1 Digestion and absorption of carbohydrate, fat and protein.

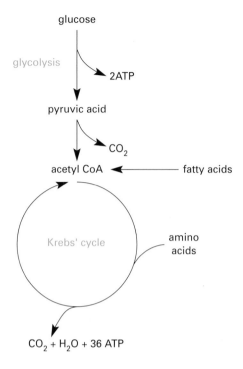

Fig. 23.2 Metabolism.

that can be metabolized are starch (from plants) and glycogen (from animal sources). Both these polysaccharides are polymers (Chapter 2) of glucose. The major polysaccharide is starch and the digestion of this begins in the mouth where *salivary amylase* begins to break it down into shorter polymers of glucose such as maltose and dextrin.

The action of amylase on starch is probably very short as food in the mouth is quickly swallowed and, when the food reaches the stomach, the *hydrochloric acid* inactivates the amylase. Some starch digestion may continue inside balls of food but, by the mixing action of the stomach movements, all the amylase is eventually inactivated.

There is no digestion of disaccharides and monosaccharides in the mouth or in the stomach. The main disaccharides which are present in the diet are sucrose, maltose and lactose (from milk). The composition of these disaccharides is given in Box 23.1. The monosaccharides in the diet are glucose and fructose.

The composition of sucrose and maltose

- **Sucrose**   Glucose–Fructose
- **Maltose**   Glucose–Glucose
- **Lactose**   Glucose–Galactose

Box 23.1

Once the contents of the stomach have passed into the duodenum the digestion of carbohydrates continues. In the duodenum, *pancreatic amylase* continues to break starch down into smaller polysaccharides and eventually into the disaccharide maltose. The intestinal juice contains three enzymes, *maltase*, *lactase* and *sucrase*, which complete the digestion of disaccharides into glucose, fructose and galactose. All the monosaccharides are subsequently absorbed in the small intestine from which they enter the blood stream and can then be metabolized in the body.

DIGESTION OF FATS (TRIGLYCERIDES)

Fats are not digested in the mouth or in the stomach, through which they pass as fat globules. The process of fat digestion begins in the duodenum where the physical properties of the fat globules is changed by the emulsifying action of bile. This breaks down the globules and increases the surface area of fat which can come in contact with the enzyme *lipase* from the pancreas. Lipase breaks fats down into fatty acids and glycerol which are absorbed from the small intestine.

In the cells of the small intestine fat is re-made and formed into small particles, with protein, called *chylomicrons*, and these are absorbed into the lymphatic system of the small intestine and taken to the blood stream. From the blood stream they are taken to the tissues of the body, e.g. liver and muscle, where they are metabolized.

DIGESTION OF PROTEINS

Protein is not digested in the mouth. In the stomach, the action of hydrochloric acid changes the physical properties of the protein, making it easier to digest. The process of protein digestion begins in the stomach under the action of the enzyme *pepsin*. Pepsin is produced in an inactive form called *pepsinogen*, which is activiated to pepsin by the action of hydrochloric acid. Pepsin breaks protein macromolecules down into smaller fragments called *peptides*.

The process of protein digestion continues in the duodenum where the enzymes *trypsin* and *chymotrypsin* also act to break the remaining protein and peptides down into smaller fragments. Both these enzymes are produced by the pancreas in inactive forms, *trypsinogen* and *chymotrypsin*, which are activated in the alkaline juice of the pancreas. This alkaline juice stops the action of pepsin.

The small intestine contains further enzymes which complete the digestion of protein down to its constituent amino acids. These enzymes are *aminopeptidase* and *carboxypeptidase* which digest single amino acids from the ends of the protein fragments. The enzyme *dipeptidase* breaks down dipeptides (composed of two amino acid molecules) into single amino acids. Amino acids are absorbed from the small intestine into the blood stream from which they are delivered to tissues for metabolism.

METABOLISM

Metabolism is composed of two processes—*anabolism* and *catabolism*—which work simultaneously to build up and break down molecules, respectively. The two 'phases' of metabolism are entirely dependent upon one another. The functions of metabolism are summarized in Box 23.2.

The two functions of metabolism

- The provision of energy in the form of ATP
- The exchange of carbon between different types of molecule

Box 23.2

All the energy-using processes of the body, such as movement, are dependent upon metabolism, as are the synthetic processes whereby large molecules such as protein are made for use in muscle and other tissues.

The Krebs' Cycle

The *Krebs' cycle* is often called the 'hub' of metabolism. The breakdown of nutrients to provide energy and the exchange of carbon for synthetic purposes is accomplished by this unique metabolic pathway.

The breakdown of glucose, and the use of glycerol in metabolism, is accomplished by the process of *glycolysis*. Glycolysis produces *pyruvic acid* and two molecules of ATP. The pyruvic acid is converted to a substance called *acetyl coenzyme A* (*acetyl CoA*) and this is further broken down in the Krebs' cycle to produce carbon dioxide, water and 36 molecules of ATP.

Fatty acids (which result from the breakdown of fats) are broken down by a process which produces acetyl CoA. The acetyl CoA is further broken down in glycolysis to produce carbon dioxide, water and ATP. The amount of ATP produced depends on the particular fatty acid which is being broken down, and it should be noted that fat produces more than twice the amount of ATP per gram as does carbohydrate and is, therefore, a much better source of energy.

Amino acids can be used to produce energy by being broken down in the Krebs' cycle. Different amino acids enter this pathway at different points.

Due to the central role of the Krebs' cycle in the metabolism of carbohydrate, fat and amino acids, it is possible to exchange carbon between molecules. For instance, some amino acids can be made from molecules in the Krebs' cycle for subsequent use in protein synthesis, and glucose can be made from some amino acids in a process called *gluconeogenesis*. Note, however, that fatty acids cannot be used to make glucose as the conversion of pyruvic acid to acetyl CoA cannot be reversed.

Chapter 24
Fluid Balance and the Urinary System

Maintaining adequate hydration and the correct distribution of fluids between the compartments of the body is vital. This chapter will look at what the normal distribution of fluid between the intracellular and extracellular compartments of the body should be. The urinary system plays a key role in ensuring that excess fluid is removed from the body. It also has an excretory function, which requires it to be constantly removing soluble toxic material from the body in the form of urine. This can be achieved even when fluid has to be conserved. This chapter looks at the anatomical structure of the urinary system and considers its ability to maintain fluid balance and respond to different fluid requirements of the body.

FLUID AND ELECTROLYTE

Water is one of the essential foodstuffs, but is a simple substance which can be absorbed and used in the body without chemical change. It enters into many of the metabolic changes which occur in the body, combining with proteins, carbohydrates and fats in digestion and being split off from them when they are used as fuel to produce energy. It is general today to consider the balance rather than the metabolism of water and salts, and the electrolytes which the salts form in the body.

Water forms the greater part of the body cells and the body fluids. Roughly two-thirds, or more accurately 60 per cent, of the body weight consists of water. This proportion must be maintained. Of this water 70 per cent is inside the body cells (intracellular) and the remaining 30 per cent is in the body fluids (extracellular); 15 to 20 per cent being in the interstitial spaces in the tissues, bathing all the body cells, even the bone cells, and the remaining 10 to 15 per cent forming the fluid of the blood, i.e. the plasma and the lymph (Fig. 24.1). These three volumes of fluid are separated only by thin semi-permeable membranes, the cell walls and the capillary walls; water constantly passes through these walls from one of these areas to the others, though the volume of each remains remarkably constant in normal health.

Fig. 24.1   Diagram to show the average percentage proportions by weight of solids and water, and the distribution of water in the body. The arrows indicate the passage of water and solutes between the three compartments containing them.

The water of which our bodies are largely made is, however, not static. Fresh water is taken in by the body each day and passes out of it by a number of channels. The total quantities taken in and passed out must balance one another. Water is taken in as water and other fluids drunk, and also in the foods eaten, which, like the body, consist largely of water. On an average the healthy man takes in 1.5 litres of fluid as water and other drinks daily, and a little over one litre in his food, an average of 2600 to 2800 ml. A similar quantity is passed out of the body by the lungs as water vapour (400 to 500 ml), by the skin as sweat (500 to 600 ml), by the kidneys as urine (1000 to 1500 ml) and in the faeces a small quantity (100 to 150 ml).

The quantities lost in urine, sweat and water vapour from the lungs vary with conditions. In hot weather and heavy work more sweat is produced to cool the body, and less urine is passed; at the same time the loss of fluid causes thirst, so that more fluid is drunk. In fever the same changes take place; there is increased loss of fluid and there must be increased intake to balance it (Fig. 24.2).

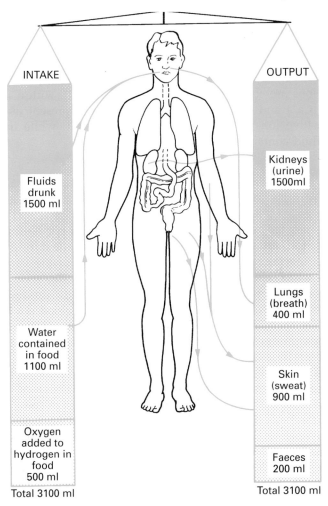

Fig. 24.2 Diagram to show how water balance is maintained by various organs.

Electrolytes

In addition to the balance of the quantity of water in the body, the body fluids must be of the right composition, i.e. they must contain the correct balance of electrolytes. These electrolytes are minute particles split off from the molecules of the various salts and dissolved in water. They are called ions (see p. 23). They carry electric charges and are of two types: negatively charged particles (anions) and postively charged particles (cations). The total of negatively charged ions must balance the total of positively charged ions. The chief negatively charged ions are chloride (Cl^-, bicarbonate (HCO_3^-, and phosphate (PO_4^-; chloride and bicarbonate are present in quantity in plasma and interstitial fluid, while intracellular fluid contains mainly phosphate. The positively charged ions are sodium (Na^+) and potassium (K^+), with a little calcium (Ca^+) and magnesium (Mg^+). Sodium is present as the main positive ion in plasma and interstitial fluid, while intracellular fluid contains chiefly potassium.

ACID–BASE BALANCE

The acid–base balance of the body fluids affects their pH. Normally the body fluids are slightly alkaline and vary very little during life, since this reaction must be correct for the action of the various enzymes which control the cell activities, such as digestion, the use of food for growth and energy production, and the production and excretion of waste products. Normally, venous blood is a little less alkaline than arterial blood because of the carbon dioxide and other acids produced in the tissues. The interstitial fluid and the intracellular fluid are similar but slightly more alkaline. This balance is maintained by the 'buffers' these fluids contain. Alkalis and proteins neutralize acids produced by tissue activities, preventing acidosis, or ketosis, a condition which leads to coma and death if the body fluids become less alkaline than normal. (The normal average is pH 7·4; see Chapter 1). In the same way, acids such as carbonate and chloride neutralize excess alkalis in the body fluids, preventing alkalosis, a condition which can prove fatal if it is not corrected. It may be caused by taking in an excess of alkaline salts such as sodium bicarbonate, or by loss of acid from excessive vomiting, or from continuous gastric aspiration after operation, or in respiratory conditions, or from retention of normal quantities of an alkali such as potassium, through renal failure.

THE URINARY SYSTEM

The urinary system consists of the components listed in Box 24.1 and shown in Fig. 24.3.

The urinary system

- The kidneys
- The ureters
- The bladder
- The urethra

Box 24.1

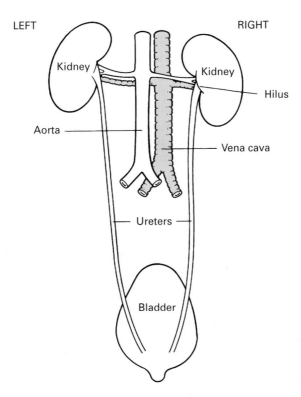

Fig. 24.3 The urinary system, viewed from the back.

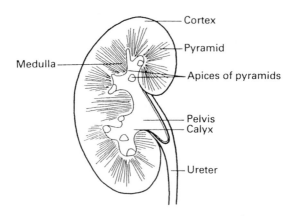

Fig. 24.4   Diagrammatic section of the kidney.

The Kidneys

The kidneys are two bean-shaped organs (Fig. 24.4) situated in the posterior part of the abdomen, one on each side of the vertebral column, behind the peritoneum. They lie at the level of the twelfth thoracic to the third lumbar vertebrae, though the right kidney is usually slightly lower than the left because of its relationship to the liver.

Each kidney is about 11 cm long, 6 cm wide and 3 cm thick and is embedded in a bed of fat called the *perirenal fat.*

The medial border of each kidney is concave in the centre. This area is called the *hilus* and it is the point at which the blood vessels, nerves and ureters enter or leave the kidney.

The kidney is enclosed in a *capsule* of fibrous tisue which can be easily stripped off. In vertical section the kidney has two distinct parts. The dark outer part is called the *cortex* and the paler inner portion the *medulla*. This leads into the collecting space which is called the *renal pelvis.*

The kidney substance consists of countless minute twisted tubules called *nephrons* (Fig. 24.5). There are over a million in each kidney. Each of the nephrons begins in a cup-shaped expansion called the *glomerular capsule*, from which the tubule leads (Fig. 24.6). Into the cup of each capsule comes a fine branch of the renal artery, forming a tuft of capillaries in close contact with its inner wall; the capillary tuft is called the *glomerulus.* The arteriole bringing blood to the tuft is called the *afferent vessel*, and the arteriole which carries the blood away is known as the *efferent vessel*; it is slightly smaller than the afferent vessel. The blood in the tuft is

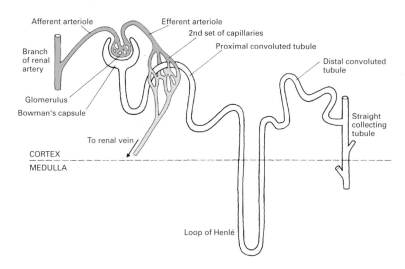

Fig. 24.5 A nephron and the blood vessels associated with it.

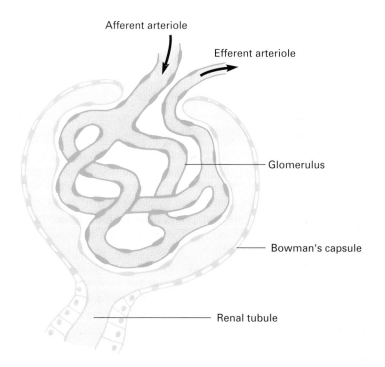

Fig. 24.6 Diagram of the glomerulus and its capsule.

under high pressure because of this and because of its nearness to the abdominal aorta.

The convoluted tubule makes a number of twists, the proximal convolutions, on leaving the capsule and then forms a long loop, the loop of Henle, which dips down into the medulla and passes back to the cortex. The tubule next makes a distal or second series of convolutions, and ultimately empties into a straight collecting tubule in the medulla.

The efferent blood vessel which comes from the capillary tuft or glomerulus in the capsule divides to form a second set of capillaries, around the walls of the convoluted tubules in the cortex. Thus the blood passes through two sets of capillaries within one organ, which does not happen in any other part of the body. The blood is collected from the second set of capillaries by small veins, which unite with other small veins to empty blood into the renal vein.

The Production of Urine

The function of the kidneys is to secrete and excrete urine. The composition of the blood must not vary beyond certain limits if the tissues are to remain healthy and this regulation depends on the removal of harmful waste products and the conservation of water and electrolytes in the body. Urine is produced by three processes (Box 24.2).

Production of Urine

- **Filtration under pressure** occurs from the glomerulus where only the thin walls of the capillaries and of the glomerular capsule separate the blood from the kidney tubule. The walls of the glomerulus are permeable to water and to other small molecules but they are not permeable to blood cells or to protein. Because the blood in the glomerulus is under pressure some of the constituents pass through into the glomerular capsule. This fluid is known as the *glomerular filtrate* and it has a composition similar to plasma in that it contains glucose, amino acids, fatty acids, salts, urea and uric acid in the same proportions. Blood cells and protein molecules are only filtered if the kidney is diseased. About 600 ml of blood per minute pass through the glomerulus and of this about 125 ml become the glomerular filtrate. If this were all excreted 150 to 180 litres of urine would be passed each day.

The average amount of urine passed each day is about 1·5 litres so it is obvious that reabsorption must occur.

- **Selective reabsorption** occurs because the cells lining the convoluted tubules are able to absorb the water, glucose, salts and their ions which the body needs. In normal health all the glucose is reabsorbed and none is excreted in the urine. Most of the water and salts are also absorbed resulting in the 1·5 litres of fluid which do pass into the collecting tubules containing normally about 2% urea. The acidity of the urine varies somewhat so that the reaction of the blood is maintained at a pH of about 7·4.

- **Active secretion** occurs because the cells lining the tubule have the ability to secrete some substances from the blood in the second capillary network into the lumen of the tubule.

Box 24.2

Reabsorption of water in the distal convoluted tubule is variable and is controlled by the secretion of anti-diuretic hormone from the posterior lobe of the pituitary gland. A decrease in secretion of ADH causes less water to be reabsorbed in the distal tubule, therefore more water is excreted as urine. The reabsorption of salts is controlled by the hormones of the adrenal cortex, especially aldosterone. The production of these hormones is increased or decreased according to the needs of the body to use water or salts and the electrolytes to which they give rise. Nervous control, together with the hormones adrenaline and noradrenaline, maintains the blood pressure at the high level required for filtration in the capillary tufts.

The *medulla* or inner portion of the kidney consists of straight collecting tubules into which the convoluted tubules of the cortex empty. It forms a number of cone-shaped masses which project into the pelvis of the kidney. These are called the *pyramids* of the medulla and are from eight to twelve in number.

The apices of the pyramids project into the pelvis and are covered with the mouths of the fine collecting tubules which pour the urine into the kidney pelvis. The function of the medulla is therefore to collect the urine secreted in the cortex and convey it to the pelvis.

The *pelvis* is an irregular branched cavity, which lies at the root or hilum of the kidney and leads like a funnel to the ureter (Fig. 24.7). Its branches, known as the *calyces* of the pelvis, penetrate into the kidney substance, and each branch receives the apex of one of the

Fig. 24.7 A pyelogram showing the pelvis of the kidneys filled with an opaque drug which has been excreted following an intravenous injection. Note that the right kidney is lower than the left one. Note also the calyces outlining the apices of the pyramids of the medulla.

pyramids of the medulla. The pyramids pour the urine into the pelvis, which conveys it to the ureters.

The Compositon of Urine
Normal urine, therefore, is formed partly by filtration under pressure from the capsules and partly by reabsorption and by secretion in the tubules. It is an amber-coloured fluid varying in colour according to its quantity. It is an acid and has a specific gravity of 1015 to 1025. (The specific gravity is the weight compared to the weight of an equal quantity of water, water being 1000).

Urine consists of water, salts and protein waste products, namely urea, uric acid and creatinine. The average composition is: water, 96 per cent; urea, 2 per cent; uric acid and salts, 2 per cent.

The percentage of urea in blood plasma is 0·04 compared to 2 percent in the urine, hence the concentration has been increased fifty times by the work of the kidney. The salts consist chiefly of sodium chloride and also of phosphates and sulphates, produced partly from the use of the phosphorus and sulphur present in protein foods. These salts must be reabsorbed or got rid of in the quantities necessary to keep the blood at its normal reaction and maintain the water and electrolyte balance. Since this reaction and salt concentration are both essential to the life of the blood corpuscles and the tissue cells this function of the kidney is very important. The normal quantity of urine secreted is 1·5 litres in 24 hours, but it is increased by drinking and by cold weather, and is decreased by reducing fluid intake and by hot weather, exercise, and by fever since these increase sweating. Potassium salts normally filter out and are reabsorbed or excreted as required, to keep the correct level in body fluids. In renal failure, their excretion may be checked so that the amount in the body fluids and tissues rises.

Countercurrent exchange

A unique mechanism is thought to operate in the kidney to ensure that, when it is required, a concentrated urine can be produced in order to conserve water. This mechanism is called *countercurrent exchange* (Fig. 24.8). By this mechanism sodium chloride (NaCl) is transported out of the ascending loop of Henlé and enters the

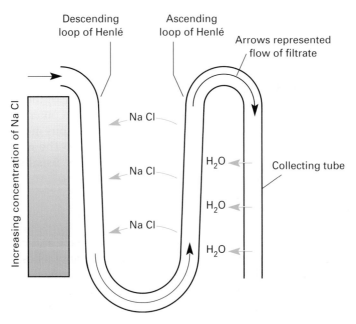

Fig. 24.8   The mechanism of countercurrent exchange.

descending loop. This leads to a concentration gradient of sodium chloride in the kidney with the highest concentration being at the base of the loop of Henlé. A similar mechanism operates in the surrounding blood vessels to ensure that, by the process of osmosis (Chapter 1) the water which is drawn into the tissue surrounding the loop is taken away in the blood stream.

The outcome of this mechanism is that the collecting tube passes through a concentration gradient of sodium chloride. When water does not need to be conserved the collecting tube is impermeable to it and a dilute urine is produced. However, when water requires to be conserved, ADH acts to make the collecting tube permeable and water is drawn into the kidneys by osmosis. The water is taken away into the blood stream by the mechanism outlined above.

The Ureters

The ureters are the two tubes which carry the urine from the kidneys to the bladder. Each is about 25 to 30 cm in length and is a thick-walled, narrow tube which is continuous with the renal pelvis and which opens into the base of the bladder. It is about 3 mm in diameter but is slightly constricted in three places; (a) at the junction with the renal pelvis; (b) where it crosses the brim of the lesser pelvis; and (c) as it passes through the wall of the bladder. These narrowed portions may be the site of impaction of a ureteric calculus (stone). The ureters, the renal pelvis and the calyces can be seen by radiography following the intravenous injection of a radiopaque substance.

The ureter has an outer fibrous coat continuous with the fibrous capsule of the kidney; a muscular coat which has an outer circular and an inner longitudinal layer, and a lining of mucous membrane which is continuous with the lining of the bladder. The muscular layer of the ureter undergoes peristaltic contractions, usually about four or five times a minute.

The Bladder

The bladder is a reservoir for urine and its size, shape and position vary with the amount of fluid it contains. When empty it lies within the lesser pelvis, but as it becomes distended with urine it expands upwards and forwards into the abdominal cavity.

Both ureters enter and the urethra leaves the bladder at its *base*. An imaginary line drawn to connect these openings outlines the area known as the *trigone*. The neck of the bladder is the lowest and most fixed part of the organ; it lies 3 to 4 cm behind the symphysis pubis. The bladder can hold over 500 ml of urine, though this would cause pain; the desire to empty the bladder is normally felt when the organ contains 250 to 300 ml of urine.

The bladder has three coats. The outer serous layer is of peritoneum, but this is found only on the superior surface. The muscular layer contains both circular and longitudinal muscle fibres; there are also two bands of oblique fibres which are situated close to the ureteric openings and which prevent urine flowing back into the ureters. The inner mucous coat is loose and is thrown into rugae when the bladder is empty. The bladder is lined with transitional epithelial tissue which allows for expansion when the organ is full.

The Urethra

The urethra extends from the internal urethral orifice in the bladder to the external urethral orifice.

In the *male* the urethra is 18 to 20 cm in length and serves as a common canal for both reproductive and urinary systems. It is divided into three portions (Box 24.3).

The male urethra

- The **prostatic portion** is about 3 cm in length and is surrounded by the prostate gland. It is lined with transitional epithelium and the *orifices of the prostatic ducts* and the *ejaculatory ducts* open into it
- The **membranous portion** is 1 to 2 cm in length and is the narrowest part of the urethra. It passes through the pelvic floor
- The **spongy portion** is about 15 cm in length and lies within the penis

Box 24.3

In the *female* the urethra is about 4 cm long and serves the urinary system only. It begins at the internal urethral orifice of the bladder and passes downwards behind the symphysis pubis embedded in the anterior wall of the vagina.

There are internal and external sphincters to the urethra; the internal sphincter is involuntary and the external sphincter is under voluntary control except in early infancy and in nerve injury or disease.

Micturition

Micturition is the passing of urine. Urine is constantly passing into the bladder from the ureters; when there is 200 to 300 ml of urine in the bladder the desire to pass urine occurs due to stimulation of the sensory nerves because of increased tension in the bladder. As the sensory impulses increase in number and frequency the motor impulses cause a reflex contraction of the bladder and relaxation of the internal sphincter. The external sphincter is controlled by the pudendal nerve. When the child has learned to inhibit spinal reflexes micturition can be delayed for a considerable time or can be induced voluntarily.

In disease micturition may be affected in many ways. The sphincter may be paralysed in a state of contraction (spastic paralysis), so that it cannot be relaxed. This will cause *retention* of the urine and the bladder will become overfull. If this is not relieved by catheterization, distension will continue and will distend the orifice which the sphincter guards, allowing a little urine to dribble away continuously though the bladder still remains full. This is called *retention with overflow*. This is bad for both bladder and kidney, causing loss of tone and interference with blood supply to the stretched bladder wall and back pressure on the kidney. It should never be allowed to occur.

The sphincter may be paralysed in a state of relaxation. In such conditions urine will dribble constantly from an empty bladder, as the bladder cannot retain it. This is comparatively rare.

In other cases of nervous disease and in anaesthesia the control of the brain over micturition may be abolished, so that the act again becomes reflex as in the lower animals; the bladder fills and empties reflexly, without the individual being aware of any sensation of fullness or being able to control the relaxation of the sphincter.

SECTION VI: PROTECTION AND REPRODUCTION

Section Introduction

Protection and reproduction are considered in the same section because they are vital for the immediate survival of individuals and the continued survival of the human race. The first line of defence, the skin, is mainly physical and it protects us by preventing dehydration and the entry of harmful foreign bodies such as bacteria. The skin also has cellular protective mechanisms and these are evident in the process of inflammation.

Protection at the cellular level takes place non-specifically, i.e. in response to invasion by any foreign body, and specifically by the development of immunity. Several cell types, which are located in a number of places in the body including the blood, the lymphatics and the skin, are involved and there is an intimate relationship between the non-specific defences and immunity.

Reproduction is possible due, essentially, to the production of different types of gametes—sperm and ova in males and females, respectively. The respective reproduction systems of males and females differ anatomically and in their physiological regulation, but both are ultimately designed to bring together sperm and ova, through sexual intercourse, in order to achieve fertilization.

Chapter 25
The Skin

This chapter covers the gross structure of the skin and the associated structures such as sweat glands and sebaceous glands. Two functions of the skin, namely temperature regulation and wound healing, are considered.

The skin covers the body and protects the deeper tissues. It contains the endings of many sensory nerves and is important in the regulation of body temperature.

STRUCTURE OF THE SKIN

The skin has two layers (Box 25.1). The *epidermis* is non-vascular and consists of stratified epithelium (Fig. 25.1). It is very thick, hard and horny on such areas as the palms of the hands and the soles of the feet and is much thinner and softer over other parts, such as the trunk and the inner sides of the limbs. The epidermis has two layers or zones: the outer is called the horny zone and the inner the germinative zone. (Fig. 25.2). The *horny zone* has three layers (Box 25.2) and the *germinative zone*, which is deeper, consists of two layers (Box 25.3).

Layers of the skin

- The **epidermis**, or outer layer
- The **dermis**

Box 25.1

The surface scales are constantly being rubbed off by friction and are constantly renewed from below as the deep cells multiply and are driven up to the surface, developing into scales as they approach it. The epidermis has no blood supply and practically no nerve supply. It is nourished by lymph from the blood vessels in the underlying dermis. It is the epidermis which is raised in a blister, and the blister can therefore be snipped with scissors without causing any pain, to allow the lymph it contains to escape.

THE SKIN

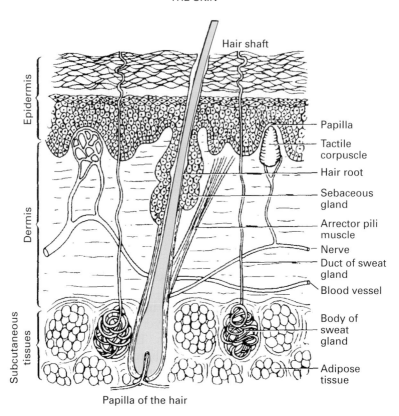

Hair shaft

Papilla

Tactile corpuscle

Hair root

Sebaceous gland

Arrector pili muscle

Nerve

Duct of sweat gland

Blood vessel

Body of sweat gland

Adipose tissue

Epidermis

Dermis

Subcutaneous tissues

Papilla of the hair

Fig. 25.1   Diagrammatic section of the skin.

Layers of the horny zone

- The **horny layer (stratum corneum)** is the most superficial layer; the cells are flat and have no nuclei and the protoplasm has been changed into a horny substance called keratin, which is water-proof
- The **clear layer (stratum lucidum)** is composed of cells with clear protoplasm and some have flattened nuclei
- The **granular layer (stratum granulosum)** is the deepest layer, it consists of several layers of cells with granular protoplasm and distinct nuclei

Box 25.2

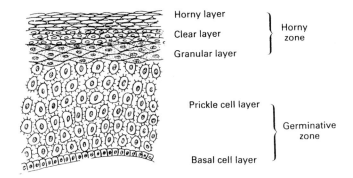

Fig. 25.2 The epidermis.

Layers of the germinative zone

- The **prickle cell layer** contains cells of varying shapes, each having short processes, rather like thorns, joining them together; the nuclei are distinct
- The **basal cell layer** consists of columnar cells arranged on a basement membrane

Box 25.3

The basal layer of cells in contact with the corium contains the pigments which give the skin its colour: yellow, white or black. The pigment protects the body from the harmful effects of the sun's rays, since dark colours absorb radiation. The scales of the surface prevent the entrance of bacteria into the tissues, since they cannot digest these dried-up cells and make their way through them. Once the epidermis is broken by a cut or prick, infection may enter the tissues and sepsis may follow.

The *corium* is a tough elastic layer which is very thick in the palms of the hands and the soles of the feet and very thin in the eyelids. It consists of connective tissue with elastic fibres, blood vessels, lymphatic vessels and nerves. Numerous conical projections, called *papillae*, extend from the surface of the corium and protrude into the epidermis. They are most numerous where the skin is most sensitive and in these areas they are arranged in parallel ridges, different in each individual, which are useful as finger prints.

The *nerve endings* in the skin are for the most part sensory, and are of different varieties to give the various different sensations of which the skin is capable, namely, the sensations of touch, heat, cold and pain. The nerves of touch end in round bodies known as the touch or *tactile corpuscles* which are stimulated by pressure, and the nerves of heat, cold and pain in delicate, tree-like branches. A few of the branches of these nerve endings pass into the epidermis. Heat is experienced only if a hot thing touches the skin over the termination of a special nerve ending affected by heat. In some parts the nerve endings are so close together that this is not appreciated, but where the nerve endings are less numerous, as over the back of the hand, it is possible to find spots where heat can be felt (hot spots) and others where cold is felt (cold spots).

The arteries supplying the skin form a network in the subcutaneous tissue and branches supply the sweat glands and the hair follicles. Fine capillaries also pass into the papillae.

Appendages of the Skin

The skin carries four appendages (Box 25.4). The *sweat glands* are twisted, tubular glands which lie deep in the true skin. Their ducts open in the pores of the epidermis, and the tube is coiled up into a little round ball deep in the skin, called the body of the gland. The sweat glands secrete sweat, which consists of water, salts, and a trace of other waste products. Much of this sweat evaporates immediately it reaches the skin surface, and is called *insensible sweat*. When sweating is heavy some of the sweat is poured out on to the skin, which becomes wet with sweat; this is called *sensible sweat*, and evaporation takes place from the whole surface. If sweating is profuse and runs off the body the cooling effect is lost. The secretion of sweat is a means of excretion of waste products, and some poisons and drugs may be got rid of in this manner; its chief importance, however, lies in the fact that the evaporation of sweat uses up body heat, since heat is required to turn water into water vapour. The amount of sweat secreted therefore depends on the amount of heat that the body needs to lose. The average amount excreted in 24 hours is 500 to 600 ml. In hot weather and in violent exercise sweating is heavy, so that much heat is lost by the evaporation of sweat. At these times less urine is passed, so that there is not excessive loss of fluid. In cold weather or rest, less sweat is secreted, so that there is less loss of heat by evaporation; at the same time more urine is secreted by the kidney, so that loss of water is balanced.

Appendages of the skin

- Sweat glands
- Hairs
- Nails
- Sebaceous glands

Box 25.4

Sweat glands are present all over the body, but are larger and more numerous in certain parts, such as the palm of the hand, the sole of the foot, the axillae, groins and forehead.

The *hairs* consist of modified epithelium. They grow from tiny pits in the skin, known as hair follicles. At the base of each follicle is a group of epithelial cells which form the root from which the hair grows. The hair root is the part of the hair within the follicle, extending through both dermis and epidermis. The hair shaft projects beyond the epidermis. The hair bulb is the expanded part of the hair within the follicle. At its base is a small conical projection called a papilla which contains blood vessels and nerves to supply the hair. Hairs are always set obliquely in the skin. The arrectores pilorum are small involuntary muscles connected to the hair follicles. They are always on the side toward which the hair leans so that when they contract the hair stands erect. The skin around the hair is elevated at the same time, which produces the effect known as 'goose-flesh'. The hairs are constantly being shed and renewed. As long as the root is healthy, fresh hair will grow from it, but if the root is destroyed or its blood supply interfered with, the growth of the hair will stop. On the scalp baldness results. Good brushing, which stimulates the blood supply, and massaging of the scalp promotes the health and growth of the hair of the head. Hair is present all over the body except on the palms of the hands and the soles of the feet, but is so fine and sparse as a rule that it cannot be seen. It is longer and more plentiful in the eyebrows, axillae, and groins, where it entangles the sweat, preventing it from running down and assisting its evaporation.

The *nails* are horny plates of modified epithelium which protect the tips of the digits. They grow from a root of typical soft epithelial cells at the base of the nail. This root is embedded in the fold of the epidermis. The nails replace the epidermis here, but are continuous with it, so that there is a continuous barrier to exclude bacteria. It is, however, easy for a break in the continuity to occur here if the

nails are not carefully looked after and this is particularly dangerous among nurses, since they are in contact with infection in the
course of their work. In animals the nails are protective in more
senses than one, but as man has invented implements of various
kinds for his use, his nails do not get worn down as do those of the
animal, and need regular cutting.

The *sebaceous glands* are small saccular glands which secrete an
oily substance called *sebum*. They are situated in the angle between
the hair follicle and the arrector pili muscle, so that contraction of
the muscle has the effect of squeezing sebum from the gland. This
lubricates the skin and hair, and keeps them soft and pliant, so that
they do not readily break. However, sebum picks up dust and
bacteria, which cling to an oily surface. As a result we must
constantly remove it by washing with soap and water. If some
substitute is not applied the skin quickly becomes brittle, so that
it breaks readily and lets bacteria enter. The nurse, whose hands are
often in lotions and soapy water, which remove the sebum, needs
to pay particular attention to this point.

FUNCTIONS OF THE SKIN

The functions of the skin are listed in Box 25.5.

Functions of the skin

- It regulates body temperature
- It secretes waste products
- It is the organ of touch and other senses through which we
  are made aware of our environment
- It keeps out bacteria by its dry, scaly outer surface
- It secretes sebum
- It protects the body by its pigment from the harmful effect of
  the sun's rays
- It produces vitamin D through the action of ultraviolet rays on
  the ergosterol it contains

Box 25.5

Control of Body Temperature

Body temperature is a balance between heat gained and heat lost. Man is a warm-blooded animal and the temperature must be maintained around 37°C; a rise or fall of one degree or more affects the normal functioning of the nervous system and of the enzymes.

The main temperature regulating mechanism is in the hypothalamus. It acts on a 'negative feedback' system; if the body temperature rises mechanisms come into action so that heat is lost from the body; if the body temperature falls heat is conserved until the temperature approaches normal.

Heat production occurs primarily by metabolic activity. Additional heat is produced by exercise, activity, increased muscle tension, shivering and also by endocrine disorders, infections, trauma and by emotion. Heat production is lowest during sleep and highest during muscular activity. *Heat loss* occurs through the mechanisms listed in Box 25.6.

Mechanisms of heat loss

- Radiation, conduction and convection of heat from the skin
- Evaporation of sweat
- Respiration
- Excretion of urine and faeces

Box 25.6

Radiation is the transfer of heat from one object to another without physical contact between the two. The body radiates heat to every object near it and the heat loss is proportional to the surface area; a large area loses a lot of heat but heat loss can be diminished by reducing the surface area. For example, less heat is lost from a body in a curled-up position than in a stretched-out position.

Conduction is the transfer of heat from one molecule to another. If one end of a metal rod is placed in the fire, heat will be conducted along it until the whole rod is hot. Direct contact will cause the body to lose heat to any object which is cooler than the body.

Convection is the transfer of heat from the body to the air, which then rises and is replaced by cooler air which in turn is heated. Heat loss by convection is reduced by wearing suitable clothing.

Evaporation of sweat from the surface of the skin is continuous and has a cooling effect on the body. It is more effective in dry climates, because when the air is humid and already saturated with water vapour further evaporation cannot occur.

Heat is also lost from the body with every exhalation of air, since expired air contains water vapour which evaporates. Only small amounts of body heat are lost through the passage of urine and faeces.

In *hot weather*, to keep the temperature normal:

1. *Heat production* is lessened; the thyroid and adrenal glands do not stimulate so much tissue activity.
2. *Heat loss* is increased by dilation of blood vessels in the skin so that radiation, conduction and convection are increased and there is increased sweating, so more heat is lost by evaporation.

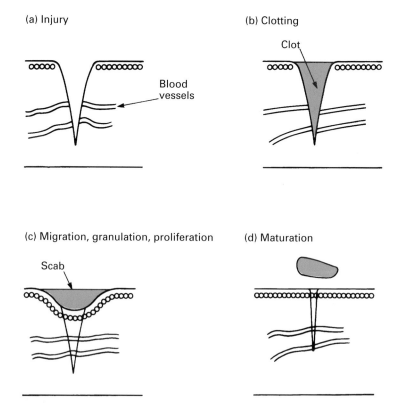

Fig. 25.3 Wound healing.

Wound Healing

When skin is damaged, for instance by a cut, it has the ability to repair itself by the process of *wound healing*. Superficial abrasions of the skin heal by re-growth of the epidermis to cover the damaged area.

If the damage extends to the blood vessels the prevention of blood loss precedes the healing of the wound, and this takes place by haemostasis involving clotting of the blood (see Chapter 15). The clot is dissolved by the action of plasmin below the surface of the wound. This allows the healing process to take place.

The process of healing (Fig. 25.3) begins with a non-specific *inflammatory response* (described in Chapter 26). The inflammation is required to bring white blood cells to the site of injury, where they remove foreign material such as bacteria which may lead to *infection*.

The next stage of healing is called the *migratory phase* during which epithelial cells migrate beneath the clot, thereby forming a *scab*. *Fibroblast cells*, which are responsible for producing *collagen* also migrate into the wound. At the same time the damaged blood vessels repair and grow in the *granulation phase*.

The *proliferative phase* is characterized by the continued growth of epithelium beneath the scab, the continued growth of blood vessels and the production of fibres of collagen by the fibroblasts. The collagen gives strength to the healing wound.

Finally, in the *maturation phase*, which may take many months depending upon the extent of the wound, collagen fibres become more organized and pull the edges of the wound together. The number of fibroblasts decreases and a normal blood supply is restored.

Chapter 26
Non-specific Resistance and the Immune System

If the physical barriers of the body such as the skin and mucous membranes are penetrated, then the body is still able to defend itself against potentially harmful material. These harmful materials, called *foreign bodies* because they are not recognized by the body as being part of itself, include bacteria, viruses and tissue from other organisms (e.g. transplants).

There are two ways in which the body defends itself: one is non-specific and will be activated whenever foreign bodies invade and one is specific—the *immune system*—and responds to specific foreign bodies. The non-specific defences are considered first.

INFLAMMATION

If the skin is cut (see Chapter 25) then chemicals are released from the damaged tissue which initiate an *inflammatory response* (Fig.

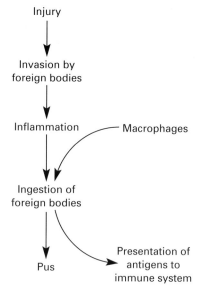

Fig. 26.1 Non-specific defences.

26.1). Inflammation can be identified by four cardinal signs (Box 26.1).

The cardinal signs of inflammation

- Redness
- Pain
- Heat
- Swelling
- Loss of function (sometimes)

Box 26.1

The outcome of the inflammatory response is that *vasodilatation* (swelling of blood vessels) takes place. This increases their permeability, and *macrophagocyte* cells migrate to the site of the injury. The macrophagocyte cells include *monocytes* and *neutrophils* and these are responsible for ingesting foreign bodies and destroying them. The inflammatory response is also necessary for the initiation of wound healing (Chapter 25). The *basophil* cells, which are called *mast cells* when they are in the skin, play a part in this inflammmatory response by releasing *histamine* which leads to inflammation.

A collection of white cells at a site of injury can be recognized as the white material *pus* which is a collection of dead cells and fluid. In some cases of injury and inflammation the body temperature can rise for prolonged periods above normal body temperature, and this is called *fever*. While a fever can be very uncomfortable and even dangerous if too prolonged, it is a response of the body which is aimed at speeding up all the processes which deal with invasion by foreign matter.

Cells of the non-specific defences, along with *T-lymphocytes*, play a role in the immune system as *antigen presenting cells* (*APCs*). By ingesting foreign bodies the APCs subsequently display parts of these foreign bodies or *antigens* on their surface, at the outside of the plasma membrane, and these are part of the process which triggers the immune system.

THE IMMUNE SYSTEM

There are two types of immunity (Box 26.2).

Types of immunity

- Cell-mediated immunity
- Antibody-mediated immunity

Box 26.2

The *T- and B-lymphocytes* respectively are responsible for cell-mediated and antibody-mediated immunity (Fig. 26.2). In response to an antigen, the lymphocytes produce T-cells which can attach to and ingest the foreign bodies, and also B-cells which produce specific protein molecules called *antibodies*, which can also attach to antigens. By attaching to the antigens they are inactivated and the resulting *immune complexes* are later ingested by the *eosinophils*.

Three types of T-lymphocyte are produced: the *T-killer* cell, which actively ingests foreign bodies; the *T-helper cell*, which helps to activate the production of antibodies; and the *T-memory cell*, which

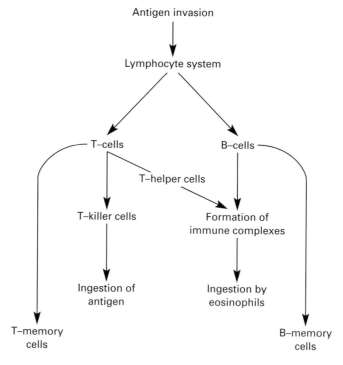

Fig. 26.2 The immune system.

is responsible for subsequent specific recognition of invasion by the antigen. Together with the *B-memory cell*, the *T-memory cell* is the basis of *immunity*.

IMMUNITY

When a foreign body is introduced into the body the immediate response is the production of a substance which will react with it and render it harmless. The foreign protein is called an antigen and substances produced in response to the antigen are called *antibodies*. Antigens can be any foreign protein, but common antigens are micro-organisms, some drugs (penicillin is an example), animal and vegetable proteins, including pollens, and foreign tissues such as transplanted organs. The reaction that occurs may be called an antigen—antibody reaction. When it occurs as a response to micro-organisms it is called immunity.

Immunity is a useful defence against infection. Each type of micro-organism which enters the body acts as an antigen and stimulates the production of specific antibody which destroys that antigen and no other. The first time a person comes in contact with the virus which causes measles, antibody is produced which enables the body to overcome the infection and which remains in the blood ready to prevent any further infection by the same type of virus. Immunity may be active, when the cells of the body make the antibody, or may be passive, when the antibody has been produced in the cells of another person (Fig. 26.3).

Active immunity may be achieved in several ways. *Active natural immunity* is gained by having the disease, after which the antibody remains in the blood ready to prevent another attack of the same disease. This type of immunity is also developed by what are called sub-clinical infections, in which the body is exposed to small

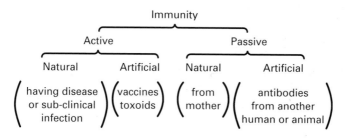

Fig. 26.3   Active and passive immunity.

numbers of micro-organisms insufficient in number to give rise to any definite symptoms, but which are sufficient to stimulate antibody production. *Active artificial immunity* is given to children and travellers in order to prevent their getting diseases which would be serious or fatal. An injection of killed micro-organisms, or live ones which have been rendered harmless, is given and the body responds by producing antibodies. In this way active immunity is built up. Harmless toxins are also used to give this type of immunity. Toxins are chemical poisons produced by micro-organisms and when rendered harmless they also act as antigens. Harmless micro-organisms are called vaccines and harmless toxins are called toxoids. Many diseases are prevented by active artificial immunity; some of the most common are whooping cough, diphtheria, measles, smallpox, poliomyelitis and tuberculosis.

Passive natural immunity is gained by the baby before birth as the antibodies are passed from the mother to the fetus. *Passive artificial immunity* is useful in the prevention of disease or in its treatment. The antibodies are produced in another human or in an animal and are injected into the person at risk. Passive immunity is always short-lived as the antibodies are destroyed after a short time.

Antigen–antibody reactions normally occur in the bloodstream and the debris is carried away by the monocyte macrophage system. When the immune reaction occurs in the tissues themselves the cells are damaged or destroyed by its side effects and this is known as *allergy*. Allergic reactions are often due to protein-like substances called *allergens*. The allergic reaction in the tissues releases *histamine* which causes redness and swelling in the skin, as in urticaria, and the outpouring of fluid as in hayfever. There may also be constriction of smooth muscle in the respiratory tract causing asthma.

The word *autoimmunity* describes the situation in which the body makes antibodies against some of its own cells. Many diseases are thought to be autoimmune in origin, two of the best known being rheumatoid arthritis and rheumatic fever.

Monocyte macrophage system

Monocytes have already been described as being one type of white blood cell which are actively phagocytic and mention was made of their being part of the monocyte macrophage system. This system is widespread throughout the body and its cells have the ability to move about which, combined with their phagocytic powers, makes them one of the most important defences of the body against micro-organisms. They are also concerned with the production of antibodies. The cells of the system occur in five parts of the body (Box 26.3).

Cells of the monocyte macrophage system

- Connective tissue, where they are called **histiocytes**
- Blood, where they are called **monocytes**
- Lining blood vessels of the bone marrow, spleen and liver and parts of the suprarenal gland and the anterior lobe of the hypophysis
- In the lymph nodes, lymph follicles of the small intestine and the tonsils
- The meninges where they are called **meningiocytes**

Box 26.3

Chapter 27
The Reproductive System

The male genital organs (Fig. 27.1) consist of six components (Box 27.1).

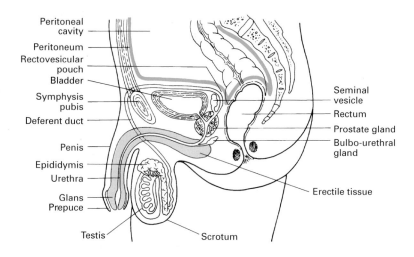

Fig. 27.1 The male reproductive organs.

The male genital organs

- The testes and epididymides
- The deferent ducts
- The seminal vesicles
- The ejaculatory ducts and the penis
- The prostate
- The bulbo-urethral glands

Box 27.1

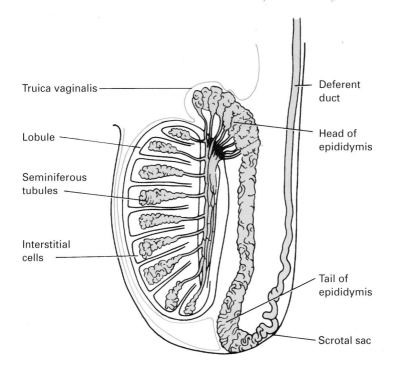

Truica vaginalis

Lobule

Seminiferous
tubules

Interstitial
cells

Deferent
duct

Head of
epididymis

Tail of
epididymis

Scrotal sac

Fig. 27.2 The structure of the testis

The *testes* (Fig. 27.2) are the reproductive glands in the male.
They are suspended in the scrotum by the spermatic cords but
they develop high up in the abdomen close to the kidneys and
gradually descend through the inguinal canal into the scrotum
shortly before birth. Occasionally one, or both, glands fail to
descend and remain in the abdomen or in the inguinal canal;
surgery may then be required to relocate them.

As the testis descends it brings down with it a pouch of peritoneum called the *tunica vaginalis*. This forms a serous covering for
the testis, but the pouch should otherwise be obliterated. If it is not
obliterated it forms a possible site for hernia. This causes an
inguinal hernia, a loop of bowel or some other organ falling into
the pouch which forms the hernia sac. Each testis consists of 200 to
300 lobules, each containing up to three tiny convoluted tubules
called the *convoluted seminiferous tubules*. The epithelial lining of
their walls contains cells which develop into spermatozoa by a
process of cell division. The tubules are supported by loose

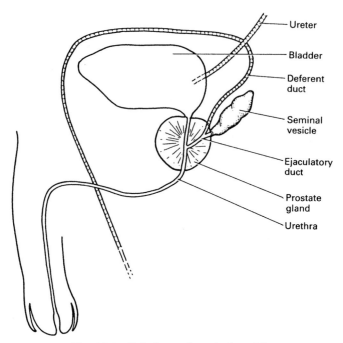

Fig. 27.3   Relations of seminal vesicles

connective tissue which contains groups of *interstitial cells*; these cells secrete the male hormone *testosterone*.

The *epididymis* is a fine, tightly coiled tube which is packed into the form of a long narrow body attached to the back of the testis. The seminiferous tubules of the testis open into it and it leads into the deferent duct.

The *deferent duct* is a continuation of the duct of the epididymis. It passes through the inguinal canal and runs between the base of the bladder and the rectum to the base of the prostate gland where it is joined by the duct of the seminal vesicle.

The *seminal vesicles* (Fig. 27.3) are two pouches lying between the base of the bladder and the rectum. They secrete an alkaline fluid containing nourishment which forms a large part of the seminal fluid.

The *ejaculatory ducts* are formed by the union of the ducts of the seminal vesicles and the deferent ducts. They commence at the base of the prostate and end at the opening of the prostatic utricle in the urethra.

The *penis* is a tubular organ plentifully supplied with large venous sinuses which can fill with blood causing erection of the organ. It contains the urethra which is common to both the urinary and the reproductive systems in the male. At the tip of the penis is an enlargement called the *glans penis*, in the centre of which lies the urinary meatus. The glans is normally covered by a loose double fold of skin called the prepuce or foreskin. It should be possible to draw the foreskin back over the glans penis, but sometimes the opening in it is too small. This is known as phimosis, and is treated either by stretching the foreskin or by circumcision, i.e. cutting away of the foreskin, in more severe cases.

The *prostate* surrounds the commencement of the urethra in the male. It is about the size of a chestnut and contains the urethra and the ejaculatory ducts. It consists partly of glandular tissue and partly of involuntary muscle and produces a secretion called semen which is alkaline and provides nourishment for the sperm.

The *bulbo-urethral glands* are situated on either side of the membranous portion of the urethra. The ducts open into the spongy portion of the urethra and the glands secrete a substance which forms part of the seminal fluid.

The *seminal fluid* is composed of substances secreted by the testes, the seminal vesicles and the prostate; it contains the spermatozoa.

The *spermatozoa* are minute cells each with a tail-like projection joined to the cell by a constricted portion called the neck (Fig. 27.4). The tail has a lashing movement which enables the cell to move after the semen leaves the male reproductive tract. As a result when the spermatozoa are deposited in the female vagina they can make their way up the uterus and uterine tubes in search of the ova. They are produced in enormous numbers and it is estimated that on average 300 000 000 are deposited in the vagina at one time, though only one is necessary to fertilize the ovum.

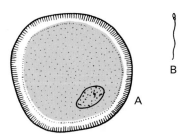

Fig. 27.4 (A) A mature ovum. (B) A spermatozöon, drawn to the same scale as (A).

THE FEMALE GENITAL ORGANS

The female genital organs consist of an internal and an external group.

Internal Organs

The internal organs (Box 27.2) are situated within the lesser pelvis.

Female internal reproductive system

- The ovaries
- The uterine tubes
- The uterus
- The vagina

Box 27.2

Fig. 27.5   The female reproductive organs, from behind.

Primary follicles

Vesicular ovarian follicle ready to burst

Ripening follicles

Ovarian stroma

Degenerating corpus luteum

Corpus luteum

Ruptured follicle

Fig. 27.6 The ovary, showing primary follicles, ripening follicles and corpus luteum.

The *ovaries* (Fig. 27.6) are two small glands about the size and shape of almonds, which are situated in the lesser pelvis, one on either side of the uterus, behind and below the uterine tubes. Each is attached to the broad ligament by a fold called the *meso-varium*. The fimbriated ends of the uterine tube and a suspensory ligament are also attached to the ovary.

After puberty the ovary has a thick cortex surrounding a very vascular medulla. At birth the cortex contains numerous *primary ovarian follicles*. After puberty some develop each month to form *vesicular ovarian follicles* (Graafian follicles) one of which usually matures and ruptures, releasing an ovum. This process is called *ovulation*. The ovum passes into the uterine tube along its fimbriated end and may be fertilized by a male sperm. If fertilization occurs it usually takes place in the lateral third of the uterine tube.

After ovulation the vesicular ovarian follicle is converted to a mass of specialized tissue called the *corpus luteum*. If fertilization occurs the corpus luteum remains active until late in the pregnancy; if fertilization does not occur the corpus luteum begins to degenerate after about 14 days. The corpus luteum produces the hormones progesterone and oestrogen; these hormones cause the lining of the uterus, the endometrium, to become thickened, ready to receive the fertilized ovum. However, if fertilization does not occur, the hormones are withdrawn as the corpus luteum degenerates and the endometrium is shed in the process called menstruation (see p. 358).

The functions of the ovary are controlled by hormones from the hypophysis: follicle-stimulating hormone and luteinizing hormone.

The ovaries begin to function at puberty and continue to discharge ova at monthly intervals from about the age of 13 to about the age of 45 years, the usual time of the menopause. The ovaries are smooth in childhood, but the scars which follow the rupture of the follicles pucker up the surface, so that they finally become very irregular and rather like almonds in appearance as well as shape.

Hormones from the ovaries are also responsible for the development of the reproductive system and the general development which marks puberty in the female, and occurs at about 13 years of age. There is marked development of the external genitals, of the uterus and the breasts: hair grows on the genitals and in the axillae; there is a general rounding of the figure and a gradual development of the typical mental outlook of the female personality, which gradually matures as the reproductive organs themselves mature.

The *uterine tubes* are situated in the upper part of the broad ligaments of the uterus. They are about 10 cm long and transmit the ova from the ovaries to the cavity of the uterus. Each tube has four parts (Box 27.3). The uterine tubes have three coats: an outer serous covering of peritoneum, a muscular coat and a lining of ciliated epithelium. The ova are conveyed along the tube by the peristaltic action of the muscular coat and by the action of the cilia.

Parts of the uterine tubes

- The **infundibulum** is a trumpet-shaped expansion which opens into the abdominal cavity close to the ovary and has a number of processes called **fimbriae**
- The **ampulla** is a thin-walled, tortuous part which forms rather more than half the tube
- The **isthmus** is round and forms about one-third of the tube
- The **uterine part** passes through the wall of the uterus and is about 1 cm in length

Box 27.3

The *uterus* is a hollow, thick-walled, muscular organ situated in the lesser pelvis between the rectum and the bladder (Fig. 27.7). It is about 7·5 cm long, 5 cm across and 2·5 cm thick and it weighs about 30 g. It communicates with the uterine tubes, which open into the upper part of the uterus, and the vagina, which leads from the lower part. The uterus forms almost a right angle with the vagina, into which the cervix of the uterus protudes. The upper part of the uterus is broad and is called the *body* of the uterus; the part of the body above the entrance of the uterine tubes is called the

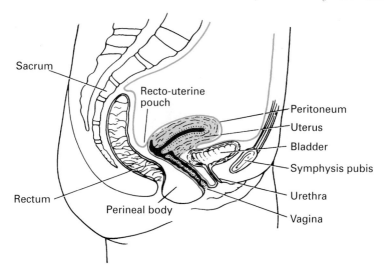

Fig. 27.7 Median section of the female pelvis, showing the position of the uterus.

fundus. The *cervix* is narrower and more cylindrical than the body and projects through the anterior vaginal wall. The narrow opening at the upper end is called the *internal os* while the lower opening is called the *external os*. The uterus has three coats; the outer serous coat is derived from the peritoneum, the muscular coat consists of involuntary muscle, the fibres of which are arranged in longitudinal, oblique, transverse and circular layers and the lining is of mucous membrane and columnar epithelium known as the *endometrium.* The *broad ligaments* pass from the uterus to the lateral walls of the pelvis. The *round ligaments* are situated between the folds of the broad ligament below the uterine tubes. The uterus normally lies in an anteverted position with the fundus facing the abdominal wall and the cervix towards the sacrum. There is also some anteflexion, the body being bent forwards on the cervix. The uterus is maintained in this position by the *transverse ligaments* which run from the cervix to the lateral walls of the pelvis and by the *utero-sacral* ligaments which run from the cervix to the sacrum. The uterus is also supported indirectly by the pelvic floor (see Fig. 14.7).

If the ovum is fertilized in the uterine tube it embeds in the thickened, vascular endometrium which has been prepared to receive it by the action of the hormones. It remains there, increasing in size, until it fills the uterus after which the uterus grows with

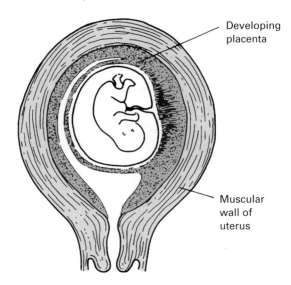

Fig. 27.8   The uterus at about the tenth week of pregnancy, showing the placenta and developing fetus.

it until the end of the period of pregnancy (Fig. 27.8). At the site of implantation the placenta develops; this is the organ through which the fetus receives nourishment and oxygen from the maternal blood during intra-uterine life.

Menstruation is the shedding of the thickened endometrium, with some blood, which occurs each month after puberty until the menopause.

The anterior lobe of the hypophysis secretes follicle-stimulating hormone (FSH) which initiates the development of a follicle in the ovary. As the ovum matures the follicle secretes oestrogen which is necessary for the growth of the endometrium and its preparation to receive the fertilized ovum. Oestrogen is also responsible for the gradual development of secondary sex characteristics in the young girl. When the amount of oestrogen in the blood reaches a high level further secretion of FSH is prevented but the anterior lobe of the hypophysis begins to release luteinizing hormone. Following ovulation (the release of the ovum from the follicle) luteinizing hormone converts the ruptured follicle into the corpus luteum which secretes the hormone progesterone. This hormone completes the development of the endometrium. If the ovum is not fertilized the corpus luteum begins to degenerate, the level of progesterone decreases and the endometrium is shed. The low progesterone level also stimulates the hypophysis to secrete more

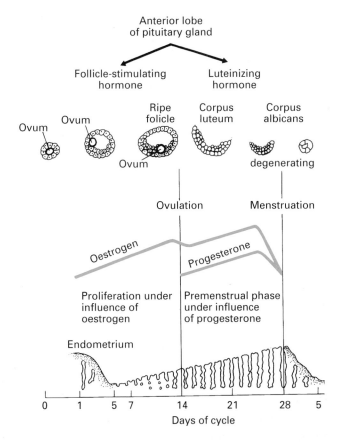

Fig. 27.9 Diagrammatic representation of the changes in the endometrium in the menstrual cycle.

FSH and the cycle begins again. For convenience of description the first day of the menstrual flow is designated Day 1 of the menstrual cycle. Ovulation usually occurs about the 14th day (Fig. 27.9).

The *vagina* extends from the uterus to the labia; it lies behind the bladder and urethra and in front of the rectum and anal canal. The cervix enters the anterior wall of the vagina at right angles so the posterior wall of the vagina is longer than the anterior wall. The recesses formed by the projection of the cervix into the vagina are called *fornices*. The walls of the vagina are normally in contact with each other. The vagina has two coats; a muscular coat which has longitudinal and circular fibres and an inner lining of mucous membrane which is in folds or rugae. After puberty this lining becomes thick and is rich in glycogen; the action of certain bacteria (Doderlein's bacilli) on the glycogen makes the vaginal secretion slightly acidic. The peritoneum passes over the body of the uterus on to the back of the cervix and close to the posterior fornix of the vagina before being reflected back in front of the rectum. This fold is called the recto-uterine pouch.

External Organs

The external genital organs in the female are collectively called the *vulva* (Box 27.4 and Fig. 27.10).

External female genital organs

- The mons pubis
- The labia majora and minora
- The clitoris
- The vestibule of the vagina
- The greater vestibular glands

Box 27.4

The *mons pubis* is a pad of fat covered with skin lying over the symphysis pubis. It bears hairs after puberty.

The *labia majora* are two folds of fatty tissue covered with skin extending backwards from the mons on either side of the vulva and disappearing into the perineum behind. They develop at puberty and are covered with hair on the outer surface after that stage. They atrophy after the menopause.

The *labia minora* are two smaller fleshy folds within the labia majora. They meet in front to form a hood-like structure called the *prepuce*, which surrounds and protects the clitoris. The labia minora unite behind in the frenulum of the labia minora (fourchette). This is merely a fold of skin which is often torn in the first labour. The labia minora are covered with modified skin rich in sweat and sebaceous glands to lubricate their surfaces.

The *clitoris* is a small sensitive organ containing erectile tissue corresponding to the male penis. It lies in front of the vulva immediately below the mons pubis and is protected by the prepuce.

The *vestibule* is the cleft between the labia minora. The orifice of the vagina and the orifice of the urethra open into it.

The *orifice of the urethra* lies at the back of the vestibule, projecting

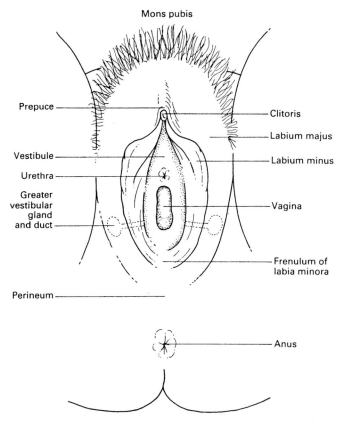

Fig. 27.10   The external female genital organs.

slightly from the normal surface level. At the entrance there are two fine tubular glands, called urethral glands, which secrete lubricating fluid and are more important because they tend to harbour infection in cases of gonorrhoea.

The *orifice of the vagina* occupies the space between the labia minora behind the vestibule. It is normally a slit from front to back, the side walls of the vagina being in contact. The orifice is largely blocked in the virgin by the *hymen*. This is a double fold of mucous membrane, usually crescent-shaped, and leaving a gap at the front for the escape of the menstrual flow. Sometimes it has a number of small perforations; this is called a fenestrated hymen. Sometimes there is no opening, but this is usually owing to the vagina not having developed into a canal. The hymen is placed a little inside the orifice of the vagina, so that there is a slight depression between the frenulum of the labia and the hymen.

The *greater vestibular glands* are two small glands lying one under each labium majus on either side of the vaginal orifice. Their ducts open laterally to the hymen. They secrete a lubricating fluid to moisten the surface of the vulva and so facilitate sexual intercourse.

The whole surface of the vulva is covered with modified skin, i.e. stratified epithelium. It carries hair on the outer surfaces only, but the inner surfaces are exceptionally rich in sebaceous and sweat glands, so that the surfaces are moist and there is no friction in walking.

The *perineum* is the expanse of skin from the vaginal orifice back to the anus. It is about 5 cm in length and bears hair. It lies over the *perineal body,* a mass of muscle and fibrous tissue separating the vagina from the rectum. The muscle of the perineal body is largely the levator ani, the chief muscle of the pelvic floor.

SEXUAL INTERCOURSE

In order for *normal fertilization* to occur, whereby male and female gametes are brought together in the fallopian tubes, *sexual intercourse* has to take place. Sexual intercourse involves the insertion of the penis into the vagina and the subsequent deposition of seminal fluid in the vagina.

In the male, sexual excitement leads to *erection* of the penis. Erection takes place when the penis becomes engorged with blood, leading to enlargement in length and girth. After erection it is possible for the male to insert the penis into the female vagina. The sensations associated with sexual intercourse lead to intense sexual excitement and *orgasm* in the male, followed by *ejaculation* of seminal fluid into the vagina.

The female responds by erection of the *clitoris*, enlargement of the breasts and erection of the nipples. Preceding and during sexual intercourse the walls of the vagina release fluid which provides lubrication. Orgasm may accompany sexual intercourse in females.

BREASTS

The breasts are two glands which secrete milk and are accessory organs to the female reproductive system. They are present in rudimentary form in the male. They lie on the front aspect of the

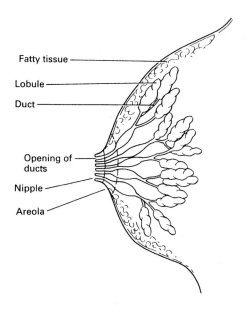

Fatty tissue

Lobule

Duct

Opening of
ducts

Nipple

Areola

Fig. 27.11 The breast.

thorax and vary considerably in size. They are circular in outline
and convex anteriorly. In the centre of the surface is the nipple,
which projects normally from the skin level and is pink in the
virgin, but pigmented after the first pregnancy (Fig. 27.11).

The breast is a compound saccular gland with its ducts converg-
ing to the nipple and opening on its surface in large numbers. The
gland tissue is similar to the tissue of the sebaceous gland, but
more highly developed and secreting milk in place of sebum. The
gland is divided into lobes by partitions of fibrous tissue—a fact
which makes the draining of an abscess in the breast difficult. The
lobes are subdivided into lobules.

The breasts develop at puberty under the influence of hormones,
and further development occurs during pregnancy as a result of
hormones from the pituitary gland and ovaries. Fluid known as
colostrum is secreted in small amounts by the gland during pregnancy and at the time of childbirth, but the true milk is not secreted
until about the third day of the puerperium, or lying-in period.

Bibliography

GENERAL BOOKS

The following is a list of textbooks in this field. Some of them are aimed particularly at nurses.

Brooker C (1993) *Human Structure and Function*, Mosby, London.
   This is a very clearly written text, specifically designed for nursing students. It has key words, learning outcomes and good summaries.
Carola R Harley JP and Noback CR (1992) *Human Anatomy and Physiology, 2nd Edition*, McGraw-Hill, New York.
   A comprehensive textbook with excellent illustrations. Particularly strong on anatomy.
Guyton AC (1991) *Textbook of Medical Physiology, 8th Edition*, Saunders, Philadelphia.
   This is a very advanced and comprehensive textbook written mainly with medical students in mind. Anatomy is minimal but as a basis to understanding disease processes this is appropriate.
Hinchliff S, Montague S & Watson R (1995) *Physiology for Nursing Practice*, Baillière Tindall, London.
   This is a new and completely revised edition of this popular, comprehensive, physiology textbook for nurses. The new edition emphasizes clinical and nursing applications and contains plentiful questions and exercises for the student.
Rutishauser S (1994) *Physiology and Anatomy: a Basis for Nursing and Health Care*, Churchill Livingstone, Edinburgh.
   Stronger on physiology than anatomy, this textbook has many interesting features to aid learning. There are some particularly helpful illustrations.
Sherwood L (1993) *Human Physiology from Cells to Systems, 2nd Edition*, West, Minneapolis.
   Purely a physiology textbook which is written at a level which is good for students who have a poor background in the subject.
Tortora GJ & Grabowski SR (1993) *Principles of Anatomy and Physiology, 7th Edition*, Harper & Row, New York.
   An absolutely comprehensive and well illustrated textbook. While it is not specifically written for nurses it does contain clinical points.
Williams P L, Warwick R, Dyson M & Bannister L H (1989) *Gray's Anatomy, 37th Edition*, Churchill Livingstone, Edinburgh.
   Probably the most comprehensive textbook on anatomy in the world and certainly the most famous. All arguments in anatomy can be solved with reference to this book.

SPECIALIST TEXTS

The following short list indicates some recommended and popular titles on issues which have been mentioned in this text.

Cummings MR (1994) *Human Heredity: Principles and Issues, 3rd Edition*, West, St Paul.
   A beautifully illustrated book which introduces the subject of genetics in a very entertaining and interesting way.
Dwyer J (1993) *The Body at War: How our Immune System Works*, Dent, London.
   A book mainly aimed at lay readers but which covers everything from microbes to AIDS.
Grace B, Davis J & Martin S (1988) *Applied Nutrition and Diet Therapy*, Saunders, Philadelphia.
   This is a comprehensive, detailed and well illustrated book on nutrition.
Morrisson (1992), *A Colour Guide to the Nursing Management of Wounds*, Wolfe, London.
   This highly illustrated text is a practical guide to all aspects of wound management.
Stucke (1992) *Microbiology for Nurses, 7th Edition*, Bailliere Tindall, London.
   Also in the Nurses' Aids Series, this is a comprehensive introduction to microbiology and infection control.

Index